PROTECTED AREA MANAGEMENT

Edited by **Barbara Sladonja**

Protected Area Management
http://dx.doi.org/10.5772/1879
Edited by Barbara Sladonja

Contributors

Joel Heinen, David Rodríguez-Rodríguez, Jafari R. Kideghesho, Tuli S. Msuya, Jelena Tomićević, Ivana Bjedov, Ivana Gudurić, Dragica Obratov-Petković, Margaret A. Shannon, Ivan Martinić, Barbara Sladonja, Elvis Zahtila, Natalia López-Mosquera, Mercedes Sánchez, Ramo Barrena, Michael Getzner, Michael Jungmeier, Bernd Pfleger, Renate Mayer, Claudia Plank, Bettina Plank, Andreas Bohner, Veronica Sărățeanu, Ionel Samfira, Alexandru Moisuc, Hanns Kirchmeir, Tobias Köstl, Denise Zak, Zoltán Árgay, Henrietta Dósa, Attila Gazda, Bertalan Balczó, Ditta Greguss, Botond Bakó, András Schmidt, Péter Szinai, Imre Petróczi, Róbert Benedek Sallai, Zsófia Fábián, Daniel Kreiner, Petra Sterl, Massimiliano Costa, Radojica Gavrilovic, Danka Randjic, Viorica Bîscă, Georgeta Ivanov, Fănica Başcău, Małgorzata Grodzińska-Jurczak, Marianna Strzelecka, Sristi Kamal, Justyna Gutowska, Isabel Mendes

Published by InTech

Janeza Trdine 9, 51000 Rijeka, Croatia

Publishing Process Manager Natalia Reinic
Typesetting InTech Prepress, Novi Sad
Cover InTech Design team

Additional hard copies can be obtained from orders@intechopen.com

Protected Area Management, Edited by Barbara Sladonja
p. cm.
ISBN 978-953-51-0697-5

For EU product safety concerns:
IN TECH d.o.o., Prolaz Marije Krucifikse Kozulić 3, 51000 Rijeka, Croatia,
info@intechopen.com or visit our website at intechopen.com.

Meet the editor

Dr. Sladonja graduated biology from the Faculty of Science, University of Zagreb (Croatia). She holds a Ph.D. in nature sciences obtained from the University consortium of Firenze, Pisa and Udine (Italy). She worked at Universities in Italy, Israel, Turkey and Malta as a researcher and specialized in Integrated Coastal Zone Management. She has been a leader on several scientific projects and published more then 70 scientific publications.
She is currently employed at the Institute of Agriculture and Tourism in Poreč. She is also active in protected areas management in the public institution for the management of protected areas Natura Histrica, Croatia.
Her recent research interests are participatory conservation and ecologic agriculture. One of her major concerns other than research is the education of new generations in environmental issues.

Contents

Preface

Protected areas are considered essential for biodiversity conservation and the main hope we have of halting the extinction of many threatened or endangered species. Since protected areas can also be a significant source of financial benefits their management is complicated and dependent on many external factors. It has been demonstrated that careful economic and social development are preconditions for successful protected area management.

This publication presents a balanced scientific consideration of the link between social and economic use of protected areas and biodiversity conservation. It reviews the experience and current status of protected area management in 12 European, North American and African countries.

The first chapter gives an overview of the historical context of the protected areas policy and international instruments related to conservation in USA. The second chapter presents new issues in protected area management and a case study in Spain. The third chapter describes the challenges of environmental protection in Tanzania, Africa. The authors provide unquestionable sound proof that global economic recession, climate change, the HIV/AIDS pandemic and political instability are potential factors, among many others, that undermine the efforts geared towards the management of protected areas. It is imperative that these issues are accorded adequate priority by mainstreaming them into policies and management plans of the protected areas and conservation agencies. This example confirms that a stable social context is essential for efficient and sustainable nature protection since it is unlikely that humans with an uncertain future and existential problems would be able to concentrate on environmental issues. Two subsequent chapters contribute to a better understanding of the relationship of tourism and the recreational values of parks, nature protection and the determination of natural values of the protected areas as well as of development perspectives of protected areas in Serbia and Croatia.

Protected areas will only be able to significantly contribute to biodiversity conservation if they are managed effectively. Standardized assessments of management effectiveness have become a powerful tool to support adaptive and effective management of protected areas over time. Chapters six and seven explain models for the evaluation of effective management, based on examples from Spain and

Austria. The authors attempt to elucidate whether protected area management effectiveness evaluations really contribute to good governance and social learning or the evaluation process itself is of greater importance.

Another two chapters are focused on Natura 2000 implementation. These chapters present the results of projects aiming at improving the management and organization of some Natura 2000 sites, with particular focus on protected areas in Poland, Austria, Hungary, Greece, Italy, Bulgaria, Romania and Serbia.

Finally, an example from Portugal describes how local community cooperation is one of the most important issues for the sustainability and development of protected areas and presents methods for evaluating the economic value of non market values.

The chapters give theoretical approaches as well as case studies from different countries, they examine possible solutions to problems and provide extensive references at the end of each chapter. The book is aimed at scientific researchers and practitioners in the field of protected area management but it is also highly recommended for professionals due to many chapters describing case studies and giving concrete solutions as well as recommending implementable measures and strategies. After all, in the words of professor Heinen (Chapter 1), I also hope we'll become smart enough to cope with the functional natural world.

The book has come to fruition thanks to the efforts and expertise of the contributing authors, as well as of good friends and colleagues. I hope that this shared effort will be the start of more collaboration possibilities in the future.

Barbara Sladonja
Institute of Agriculture and Tourism Poreč
Croatia

Chapter 1

International Trends in Protected Areas Policy and Management

Joel Heinen

Additional information is available at the end of the chapter

http://dx.doi.org/10.5772/50061

1. Introduction

Traditional human societies have protected natural areas for various cultural purposes for millennia. Examples include the sacred forests of South Asia and parts of Africa, sacred burial grounds of some native American groups and traditional royal hunting reserves in many parts of Europe, Asia and Africa, which were generally only seasonally opened for hunting by royalty (Borgerhoff Mulder & Coppolillo, 2006). The modern concept of the national ownership and protection of natural areas for the benefit of society at large is a much more recent phenomenon; the United States became the first country to conserve nationally protected areas with the creation of Yellowstone National Park in 1872. This Category II (below) protected area is now close to 9,000 square kilometers in size and was inscribed as a World Heritage Natural Site in 1978.

The management of many of the earliest protected areas would be at odds with modern conservation practices. For example, for several decades after its creation, the US Calvary managed Yellowstone and mounted soldiers regularly hunted bison and elk for food - and wolves as vermin - within its borders. By the 1930s, wolves had been eradicated from the park, and remained absent until the mid-1990s when the US Park Service and US Fish and Wildlife Service jointly reintroduced the species from animals captured in Canada. Within several decades of the creation of Yellowstone National Park, Canada, New Zealand and Australia all had set aside protected areas and had begun developing national legislation to manage them, and the United States began establishing wildlife refuges as a separate category of protected area (Fischman, 2003).

Much has been written about the historical and cultural context of this (then) new phenomenon, and the similarities of its earliest adherents. All were British colonies, spoke English as their national language, and were being quickly populated by immigrating Europeans. All four countries also had *de facto* policies of subduing their native peoples to the point of what many now consider cultural genocide. This had the effect of depopulating large natural areas, within even larger countries with low population densities to begin with, in a rather short time period during the late 19th and early 20th Centuries. Some historians also note that the European Diaspora naturally tended to look to Europe for its cultural inspiration. The countries of the Old World had great universities, museums, artworks, palaces and ruins dating back to ancient Greece and Rome, while the New World had scenery and natural areas unsurpassed by anything in densely populated Europe. This school of thought considers the movement to create national protected areas to be motivated, at least in part, by 'Europe envy' (Zaslowsky and Watkins, 1986). By the early 20th Century, all four of those countries and a few others (*e.g.* Sweden) had set aside multiple natural areas and had created professional management authorities to protect them. Canada was the first country to create a national park management agency (in 1911) followed by the USA (in 1916).

In any case, there is much evidence to suggest to that the earliest parks (and many still; see below) were not set aside with particular reference to conservation in any form, and thus the 'Europe Envy' thesis is generally accepted. Most of the earliest units contained spectacular scenery, but their borders did not relate to the habitat needs of native species, much less the dynamics of entire ecosystems (Norton, 2005). By the 1930s, the American park system received criticism from within with a report by Dixon and Wright, two Federal employees, that received widespread attention. The authors stated that most units were "mountain top parks" and preserved only scenery with no regard for wildlife. Seasonal movements of many species were such that large populations of birds and mammals were left outside park boundaries (and therefore subjected to hunting); the early American 'mountain top parks' were, therefore, ineffective for many conservation purposes (Dombeck & Williams, 2003).

Modern conservation biology has also greatly expanded our ideas of the geometric design and placement of protected areas across landscapes (Primack, 2006), but the problem of 'mountain top parks' still remains. For national governments, it is simply easier to set aside large protected areas in places such as high elevations, deserts, tundra, *etc.*, *i.e.* where there are few competing economic demands, than in areas of high biological productivity. The latter tend to be at low elevations, in temperate or subtropical zones, and receive adequate rainfall. In short, the most productive ecosystems are also those where humans tend to concentrate. Tropical rainforests, with their primary productivity largely found in the canopy and frequently harboring human diseases, are possible exceptions to this generality, but in that cases too, they are at risk worldwide (Wilson, 1999).

Canada and the United States also pioneered several other conservation movements during the Progressive era of the early 20th Century. They developed the world's first international treaties on the protection of migratory wildlife, with separate instruments for wild salmon, fur seals and migratory birds (Dorsey, 1998). The last, The Migratory Bird Treaty Act of 1918, is still in force. Canada and the United States also developed the world's first transboundary protected area in 1932, with the creation of Waterton-Glacier International Peace Park, a large area that conserves habitat on both sides of the international border in the northern Rocky Mountains. With this came the recognition that many species and ecosystems cannot be conserved within the borders of single nations and these legal instruments were watershed events in the history of conservation worldwide (Susskind, 1994). From these humble beginnings, many other bilateral, regional and global conservation conventions have been developed for the protection of both migratory species and natural areas.

The largely Western ideal of protected areas as raw nature devoid of humans (except for tourism) was never really true to begin with; most areas set aside in the nations that began the movement had been occupied by pre-industrial people who were removed. This concept was also largely out of synch with realities on the ground in developing nations During the post World War II period of decolonization, many seminal wildlife studies were conducted in various places in Africa and Asia and the world became much more aware of their unique heritage. The International Union for the Conservation of Nature and Natural Resources (now IUCN – The World Conservation Union; www.iucn.org) was begun in 1948 with a charter to develop world wide standards for conservation and the World Wildlife Fund (www.worldwildlife.org; now the Worldwide Fund for Nature) was established several years later, initially as a fund raising mechanism for IUCN. Having been developed in the West, with essentially all funding coming from West, meant that Western standards of nature conservation were becoming global (Swanson, 1997). IUCN's World Commission on Protected Areas (WCPA) was organized in the 1950s, and developed internationally-recognized categories of protected areas by the 1970s, which were modified in the 1990s (see below).

Post-colonial governments in developing nations began setting aside protected areas by the 1960s, but the 'fences and fines' approach of the West had its limits in this context. Some, such as Kenya, Tanzania and India, already had the semblance of a protected area system as a result of colonial British rule, but these were areas largely set aside for use by British government officials and indigenous elites for hunting reserves, and effectively prohibited rural residents, who were dependent on natural resources, from entry (*e.g.* Gillingham & Lee, 2003; Bruyere *et al.* 2009). In 1962 and 1972, IUCN held its First and Second World Conferences on Protected Areas, respectively. Both were characterized by over-representation of delegates from developed countries and there was little focus on the issues relevant for newly emerged developing countries. This began to change with the Third World Conference on Protected Areas, held in 1982, in Bali, Indonesia. The Conference was renamed "National Parks, Conservation and People" and the theme was the role of protected areas in economic development; a majority of participants came from developing countries. The Fourth and Fifth World Conferences were held in Venezuela (1992) and South Africa (2002) respectively, and the global agenda for protected areas in each decade expanded from the one preceding it.

The relative success of national parks in the United States, Canada, Australia and New Zealand was due at least in part to the fact that population densities were low in those countries to begin with and that indigenous peoples had been largely removed from many ancestral areas as part of national policy as those countries were developing. Such was not the case in the developing world, and there is now near universal agreement that the Western national park model is generally inappropriate for the situation in most developing countries with their large rural populations dependent (at least in part) on extractive activities in natural areas (*e.g.* Campbell & Vainio-Mattila, 2003; De Boer & Baquete, 1998; Groom & Harris, 2008; McShane and Wells, 2004)). The WCPA recognized this with the liberalization of rules regarding national parks and more strictly protected areas, and with the modification of protected area categories recognized worldwide in 1994 (below).

2. IUCN – WCPA categories of protected areas

IUCN defines a protected area as "an area of land and/or sea especially dedicated to the protection and maintenance of biological diversity, and of natural and associated cultural resources, and managed through legal or other effective means." According to the World Database on Protected Areas compiled by the WCPA, there were over 7,000 separate units covering over 17,000,000 square kilometers as of 2007. This includes about 3.3% of Earth's total surface area and nearly 10 % of Earth's land surface, but less than 0.5% of its sea surface, although there has been recent growth in the designation of near-shore marine protected areas as well. The WCPA's mission is to "promote the establishment and effective management of a world-wide representative network of terrestrial and marine protected areas as an integral contribution to IUCN's mission." A general goal is to bring 10% of the Earth's land surface, including 10% of all recognized ecosystem types, under one or another internationally recognized category of protected area. The growth of such areas has been very rapid during the past several decades, but, based on the aforementioned criteria, some ecosystem types are over-represented, while most, and especially the more productive ones, are under-represented (Chape *et al.* 2008).

The WCPA uses a system in place since 1994 to define these areas (Table 1). Here I describe the major management categories, but please note that many nations have additional protected natural areas that do not fit within the IUCN criteria and are thus not included on the United Nations List of Protected Areas. Based on IUCN criteria, national protected areas are those managed by the "highest competent authority'' which, in most cases, is the national government. Yet many countries have State, County, Provincial or Urban parks, recreation areas, *etc.*, in additional to those designated at the national level. In some cases, depending on the management plan, size and remoteness of such areas, they are included on the World List, but in many other cases they are not.

Similarly, many countries have private reserves (*e.g.* Nature Conservancy reserves, land trusts, *etc.* in the United States and elsewhere) or reserves managed by other entities (*e.g.* university-owned research reserves), that are generally not included based on IUCN criteria. In many cases, national forests or rangelands, which can be important for habitat for many native species, are also not included because their permitted uses exceed that considered appropriate by IUCN. With these caveats in mind, it is generally true that there is much more natural area set aside (about 17% of the Earth's land area; Chape *et al.* 2008), albeit in small reserves and/or under greater degrees of human uses, than is recognized internationally based on IUCN criteria. IUCN categories, based on a numbering system from most to least strictly protected, are as follows (from www.iucn.org/about/union/commissions/wcpa):

Ia. Strict Nature/Scientific Reserve. The main purposes of Category Ia reserves are scientific research and species conservation, and other human uses are generally banned. Because of this, few nations recognize Category Ia reserves within national law, but quite a few have *de facto* Strict Nature reserves. These may include, for example, very remote regions of much larger protected areas in which inaccessibility precludes tourism or other uses.

1b. Wilderness Areas. Wilderness areas are generally large and remote. Tourism is permitted, but since permanent human dwellings and motorable roads generally are not, tourist numbers are few and generally involve backpacking style camping. They provide for the protection of wilderness and maintenance of ecosystem services. This category was added in the WCPA category revisions of 1994.

	Management Objectives							
Category and Name	SR	WP	SD	ES	NF	TR	ED	SU
1a: Strict Nature Reserve	***	**	***	**	-	-	-	-
1b: Wilderness Area	*	***	**	***	-	**	-	-
II: National Park	**	**	***	***	**	***	**	*
III: National Natural Monument	**	*	***	-	***	***	**	-
IV: Habitat or Species Management Area	*	*	***	***	*	*	**	**
V.: Protected Landscape Or Seascape	**	-	**	**	o**	***	**	**
VI: Managed Resource or Extractive Reserve	*	**	***	***	*	*	*	***

Table 1. IUCN -The World Conservation Union Protected Area Management Categories (adapted from IUCN 2003)
Key to Management Objectives: SR, scientific research; WP, wilderness protection; SD, species or genetic diversity conservation; ES, environmental services; NF, natural or cultural features; TR, tourism and recreation; ED, education; SU, sustainable use; and CA, cultural attributes.
Key to importance of objectives by category: *** designates a primary objective; **, a secondary objective; *, potentially not applicable; and -, not applicable.

Category II. National Parks. This has been the most used protected area category worldwide. National parks are generally large areas that protect more than one important natural feature and/or wildlife population, and in which tourism is generally permitted and promoted. Other important functions include providing environmental services and opportunities for environmental education as well as scientific research. National parks tend to be the best known and most important protected areas economically, and many of the best examples worldwide are also recognized internationally as World Heritage Sites.

Category III. National (Natural) Monuments. This category has, in general, the same aims as Category II, but national monuments are generally smaller than national parks and are set aside to protect one or several important natural features. In some cases, these can be combined with cultural features in a natural setting (*e.g.* Mt. Rushmore in the United States). Because of their generally smaller size, they are usually not important for broader conservation purposes such as ecosystem services, but many contain important wildlife populations.

Category IV. Managed Habitat/Wildlife Reserves. By sheer numbers, this is the second most important protected area category worldwide. In general, these reserves are established to protect one or more important wildlife populations and, for the larger units, they can also be important for providing ecosystem services. Tourism is frequently permitted within them, but not promoted as in the case of II and III, above. In addition, material alteration can take place within Category IV protected areas to enhance habitat for the species of conservation concern. For example, maintaining pastures for ungulate grazing, creating empoundments for waterfowl habitat, *etc.*, may all be permitted within them. These activities are generally not permitted in the previous categories. Sustainable use is frequently a secondary goal of Category IV reserves, and some (limited) hunting of common game species may be permitted within some, or in adjacent areas.

Category V. Protected Landscapes/Seascapes. Category V reserves are perhaps the most interesting for their breadth of permitted activities and management options. These are generally large areas set aside for a combination of their natural and cultural features, and they generally promote tourism. In many places, human habitations are found within them, including small towns with examples of rural working landscapes. As such they are generally designated across landscapes that contain an admixture of public, semi-public and private lands, and may be quite altered from their natural state.

Category VI: Managed Resource/Extractive Reserves. Category VI, like Ia, was added to the list of protected are categories with the 1994 revisions. These are generally large reserves that provide for ecosystems services, but their main purpose is the conservation and sustainable use of important species and their gene pools. Active removal of forest products is permitted and in fact encouraged, and, as such, they tend to be important economically for local communities. The general rule for a protected area to qualify under this category is that no more than one third the area can be subject to intense harvest. Many countries (the United States included) have large areas set aside in which more extensive harvesting is permitted. Such areas may be managed in a semi-natural state for national purposes, but do not qualify as Category VI internationally (*e.g.* National Forests in many countries).

3. Some caveats of protected area categories

In the modern era (post 1990), greater areas under Category V and VI reserves, both terrestrial and marine, have been created worldwide than other reserve types. This is especially true in developing countries, but is also true in some of the large marine protected areas created in the United States (*e.g.* the Florida Keys National Marine Sanctuary). Given their more lenient management regimes, this is also not surprising due to the dependence in many places that rural residents have on natural resource extraction and use. However, there has been a great deal of concern expressed, especially by natural scientists, about this phenomenon. Since large predators, especially, are generally not tolerated by humans (and vice versa), and yet are keystone species in many ecosystems, Category V and VI reserves are especially problematic from a purely ecological standpoint (*e.g.* Heinen and Mehta, 1999). Yet these reserves can be the most important from a purely economic standpoint (*e.g.* Sherman and Dixon 1990) and from the standpoint of human cultural values.

While debates were ongoing in the western academic literature largely between natural and social sciences on the competing values of different types of protected areas and their uses, with ecologists generally favoring more strict protection and social scientists favoring less strict protection (*e.g.* Redford and Sanderson, 2000), many nations, as well as less philosophically-driven researchers and development workers, were slowly arriving at a different consensus. That is, both sides have valid arguments and large reserves, and the regions in which they are found, can have elements meeting these competing demands via zoning criteria. For example, India and Nepal added less strict regulations to some of their national parks (including some limited extractive uses), while keeping more strict regulations in others, and both also actively supported buffer zone policies in the vicinity of more strictly protected areas beginning in the 1990s (*e.g.* Heinen and Shrestha, 2006). In those cases, many Category II and IV protected areas are surrounded by buffer zones that are managed more like Category V or VI protected areas, whether or not they are recognized as such internationally.

To promote the broad goals of sustainable development as articulated in the 1987 Brundtland Report (Bruntland, 1987), Agenda 21 (Sitarz, 1993) and the 1992 Convention on Biological Diversity (Glowka *et al.*, 1994), rather intensive local development inputs are needed in such areas to reduce demands on the core protected area (*i.e.* the Category IV or lower reserve). This may include rural enterprise development such as farm fisheries, agro- and community forestry and training of local people for tourism related jobs. There is now a large and growing literature on the development and success of community-based conservation (CBC) programs and integrated conservation and development programs (ICDPs) that is generally outside the scope here. Suffice it is to say that, for our purposes, from both socio-economic and ecological standpoints, there is growing evidence that this mixed approach has many advantages and pitfalls (*e.g.* Fiallo & Jacobson, 1995; Lepp & Holland, 2006; Lepp, 2007).

Some ecological factors that may lead to success (or not) include the types of species protected in core areas (*e.g.* large mammals frequently cause much loss to local farmers, including lost human lives on occasion) and the types of natural plant products and other resources (*e.g.* fish), their growth and sustainable harvest rates and local market values, that may be harvested legally from designated extractive zones. Socio-economic and other factors are many and varied. For example, human population density alone, and especially the ethnic heterogeneity and recency of immigration to an area, can determine the degree of difficulty of developing and sustaining CBD programs (Heinen, 1996). Recent research has shown that the creation of protected areas and development inputs into their surroundings can act as attractants for new immigrants, further complicating the issue (Wittemyer *et al.* 2008). In addition, increased wealth (due to tourism and other employment opportunities) of residents around protected areas can also create difficult managerial consequences in their vicinity via increasing demand for many forest products (*e.g.* Fu *et al.*, 2004). But, in general, CBC programs in areas that are more stable demographically and/or especially areas in which they have been in place for longer time periods (and thus institutional trust and social capital has been built), have been shown to be effective over time in many case studies (Baral *et al.* 2007). But this can take many years to a few decades.

The protected area categories used by IUCN's WCPA are broad enough to cover quite a bit of the world's protected natural heritage adequately, but individual countries deviate from international standards frequently. As previously mentioned, they are not inclusive enough to capture many of the world's smaller protected areas (*e.g.* state, provincial and country parks) or important private reserves. Such reserves can be very important for the conservation of local plant and insect species, as stopover areas for birds during migration, as important nesting areas for species such as sea turtles, and for the ever-increasingly important purposes of introducing urban and suburban populations to environmental and science education, which are all very important objectives. For example, the Counties and the State of Florida maintain a system of such reserves in the urban and suburban matrix of Miami-Dade, Broward and Palm Beach in Southeast Florida that are heavily visited by residents and tourists (Alonzo & Heinen, 2011); their combined attendance annually is thought to be greater than for nearby Everglades National Park. As such, small reserves can be disproportionately more important for several simultaneous conservation goals than some internationally recognized large reserves.

Individual countries may also vary quite a bit in terms of management practices and hence in terms of how the WCPA categorizes their protected areas. For example, 'National Parks' under both British and Japanese standards are frequently too materially altered to be considered Category II protected areas by WCPA. Because they may include many important cultural components and have private in-holdings, and, in some cases, entire towns, they are generally classified as Category V by IUCN. Many of the large extractive reserves in the USA (and elsewhere) simply allow too much extraction to be classified as Category VI reserves (*e.g.* US National Forests managed by the Forest Service and Grazing Areas managed by the Bureau of Land Management). Other units within the same system, under less intensive management, do qualify, and thus IUCNs' WCPA must consider each unit on a case by case basis by reviewing individual management plans in deriving the United Nations List of Protected Areas to determine if each qualifies. World wide, the effort required is huge and the list is always in need of updating. In spite of these caveats, the system has proven useful for over 15 years; it is also adaptable and widely recognized, so there is little reason to change it at the present time.

Another issue that is frequently debated and studied is that of 'paper parks.' These can be defined as units that are protected at the national level via appropriate laws, and in some cases, recognized under international conventions as very important protected areas, but in which there is either inadequate or no active enforcement on the ground. This term is now applied to many parks and reserves in developing countries where inadequate budgets and manpower for conservation are the norm. The World Heritage Convention (below) has focused on this issue and has developed the "List of World Heritage in Danger" (whc.unesco.org/en/danger). Inscription of this list should cause national shame, for these are some of the most spectacular parks on Earth, but in fact, the vast majority of protected areas are not World Heritage Sites, and there are many 'paper parks' in which poaching, logging, or other extraction go on regularly in spite of laws. WCPA has no means currently to assess these issues on a case by case basis worldwide, so many listed sites (especially Category II) either should be placed in another less strict category or removed from the List.

4. International regimes concerning protected areas

There are currently dozens of international instruments related to the conservation of species, natural areas, or both. The vast majority are bilateral or regional; some are quite well studied while others are more obscure (Klemm & Shine, 1993). Regional treaties in this area exist in Europe, the Commonwealth of Independent States (*i.e.* most of the former Soviet Republics), Southeast Asia, sub-Saharan Africa and Central America. Here I briefly discuss the major international regimes that are subject to ratification or acceptance by all United Nations members, but interested readers are referred to Klemm and Shine (1993) for more information on some of the regional agreements.

4.1. The man and biosphere program

In the late 1960s, the Man and Biosphere Program (MAB) was conceived under the auspices of the United Nations Educational, Scientific and Cultural Organization (UNESCO) based in Paris (Batisse, 1982, 1986). By 1971, MAB was implemented with the broad goals studying human relationships with the biosphere, especially for studying long-term human induced impacts and conservation for sustainable development. A major objective of MAB since its beginning was to develop a worldwide network of international biosphere reserves and, and there are currently (2009) 553 MAB-designated Biosphere Reserves in 107 countries (www.portal.unesco.org).

Based on MAB criteria, biosphere reserves are established in representative ecosystems for research purposes, with several secondary goals. These include: preserving traditional forms of land use, disseminating knowledge to manage resources, and promoting cooperation in solving resource related problems. The biosphere reserve model is one in which more strictly protected core reserves are surrounded by nested buffers permitting more human uses with distances from the core. Given the time period, the goals of the program and the concept of reserves with functional buffers, were quite progressive and a number of countries have since followed suit with the zoning implicit in the MAB reserve design (above). Education and training are also promoted under MAB, as is ecosystem level management. A number of countries in Latin America and the former Soviet Union now use 'biosphere reserve' as a category of nationally-protected areas; many of these sites are listed on UNESCO's international list, while others are not.

The type and scale of reserves listed under MAB vary quite a bit based on national norms and conventions, and national MAB programs are given a great deal of leeway in nominating sites (Heinen & Vande Kopple, 2003). Any nomination is subject to acceptance by the international MAB program. MAB is also quite fluid in maintaining ties with international organizations (*e.g.* the World Wide Fund for Nature, WWF; the World Wildlife Fund in the USA), and United Nations agencies such as the United Nations Environmental and Development Programs, respectively, and with Secretariats of international conventions such as Ramsar and CBD (below). Through international as well as regional programs, MAB fosters exchanges among reserves and facilitates interactions through networks such as AfriMAB and EuroMAB.

Sites listed under MAB range from large and nationally protected areas (*e.g.* Everglades National Park) to much smaller reserves maintained by sub-national entities (*e.g.* The University of Michigan Biological Station). While some may perceive this as a weakness in that such disparity in size and purpose leads to little uniformity in management of these reserves, this can also be considered a strength of MAB. That is, individual national programs can promote the overarching goal with a number of different reserve types, and can take part in various levels of international cooperation, as long as the reserve meets the general criteria of research, education and outreach and includes some semblance of the zoning criteria in which core areas are well protected. As such the program is quite flexible and unique. The fact that it is not legally binding can also be considered both a strength and a weakness as it, again, promotes more flexibility but less uniformity. In any case, the MAB program has existed for four decades and in many ways set additional standards for protected areas management internationally. As such, it is quite important for the movement worldwide.

4.2. The ramsar convention

The Convention on Wetlands of International Importance, Especially for Waterfowl Habitat was conceived in Ramsar, Iran in 1971, and is generally known as the Ramsar Convention (Hails, 1996). Ramsar was the first truly international convention promoting the protection of natural areas and, in many ways, it remains the most important. Like MAB, it was also very progressive for its time (below). To date (2009) there are 159 contracting parties and

1,847 sites included on the List of Wetlands of International Importance, which collectively cover about 1.8 million square kilometers of area (www.ramsar.org). The purpose of Ramsar as articulated in the Preamble is to recognize the interdependence of humans and the environment and to consider the ecological and economic functions of wetlands as fundamentally important. Parties are instructed to develop national wetlands policy with the aim of decreasing wetland loss, and to recognize that waterfowl, by virtue of their annual migrations, are an important international resource. All Parties must nominate at least one Wetland of International Importance from within their borders.

Ramsar defines wetlands very broadly, to include fens, bogs, marshes and swamps as well as near-shore marine areas in which low tide does not exceed 6m in depth. In this way, the Convention was progressive in that near-shore marine areas can be included. At the time of its formulation, very few nations had created marine reserves, but this movement has increased greatly in the decades since, and many coastal areas are now listed as Ramsar sites. Similarly, Ramsar provides a very broad definition of waterfowl to include any species of migratory bird that uses wetlands for any part of its life cycle. As such, waterfowl in the traditional sense are included (*i.e.* ducks, geese and swans), as well as all species of waders and fishing birds, and a number of passerine species that breed in wetland areas.

Various Articles of Ramsar further articulate the responsibilities of Parties to conserve wetland areas. Article 4, for example, encourages Parties to create wetland reserves whether or not they are listed sites, and to train personnel for research and management of wetlands, while Article 5 instructs Parties to consult about implementing the Convention, an important provision for sites that may cross international borders. Ramsar was also historically important in promoting the concept wise use of wetland resources for sustainable development. This also made it very progressive for its time in the sense that the Bruntland Report was released 15 years after Ramsar, and the Convention on Biological Diversity followed Ramsar by 2 decades. It also preceded the changing concepts of the WCPA about protected area management categories and the promotion of sustainable human uses within more categories than had been the case previously (above). Ramsar also maintains a Trust Fund for which Parties that are developing nations can apply for funding for special projects to maintain sites, offer trainings, *etc.*

In another sense, however, Ramsar can be criticized for having relatively little control over Parties as to how they manage wetlands overall. The idea of no net loss of wetlands, inherent to Ramsar, has frequently not been met, even in the United States, and wetland drainage continues in many Party States. Listed Ramsar sites themselves vary quite a bit in terms of their importance. For example, Canada and the United States have relatively few listed sites, but all are large and of obvious importance for the broad goals of the Convention (*e.g.* Everglades National Park). While many of the smaller and much more densely populated European countries list large numbers of small sites, some of which are of dubious importance. Even with these caveats in mind, Ramsar is very important for international conservation for promoting wetland protection and for many broader issues related to protected areas management. It could also be used as a template to form other

Conventions focused on single broad ecosystem types (*e.g.* tropical forests or tropical grasslands), although none currently exist.

4.3. The world heritage convention

The International Convention for the Protection of World Cultural and Natural Heritage (The World Heritage Convention or WHC) as adopted by UNESCO in Paris in 1972. As of 2009, WHC had 186 Parties; of these, 148 have sites listed on the World Heritage List. Of the 890 sites listed worldwide, the majority are Cultural heritage sites (689) and will not be considered here (whc.unesco.org). Of the remaining, 176 are Natural heritage sites (mostly national parks) while 25 are mixed sites containing both cultural and natural heritage. WHC came into force in 1975 with the purposes of conserving both natural and cultural areas of outstanding universal importance. As such, Parties recognize that many sites are of importance to world's heritage and not just to the heritage of the countries that may contain them. In addition, to maintaining the World Heritage List, WHC's Secretariat also maintains the World Heritage Trust, under which developing countries can apply for project funds to help maintain sites. Lastly, the Secretariat also maintains a list of World Heritage Sites in Danger for previously listed sites that are under improper management or for some other reason at risk. Sadly, the site nearest my own desk, Everglades National Park, is currently on this list due to the lack of progress of the Comprehensive Everglades Restoration Project under the Bush Administration. Compliance is a major issue, and focus of study, regarding many such legal instruments (*e.g.* Faure & Lefevere, 1999).

Many of the Articles of WHC pertain solely or mainly to cultural sites but there are several important provisions that relate to natural sites. Article 2, for example, states that natural heritage consists of "physical and biological formations or groups of such formations, which are of outstanding universal value." The definition is further clarified to include areas that constitute important habitat for endangered species of universal value, outstanding geological formations, or other natural features of outstanding beauty. Parties to the Convention are responsible for proposing sites within their borders for listing, and providing strong evidence based on a set protocol for each place that allegedly constitutes a site of outstanding universal value. Most natural sites on the list were already world famous before they were listed (*e.g.* The Grand Canyon, The Serengeti, Mount Everest, The Great Barrier Reef, The Galapagos Islands), and in fact, most were already protected under national law as National Parks or other types of internally-recognized protected areas. None-the-less, there is national prestige to having sites listed as World Heritage, and the added (although rather meager) incentive for developing nations to garner some funds through the Trust. National governments and private tour operators alike frequently use listing in advertising as an incentive to attract more tourism, and World Heritage Natural Sites are among the most-visited protected areas on earth. In Nepal, for example, of 16 nationally protected areas, the two World Heritage Sites alone (Everest and Chitwan National Parks) typically account for over one third of tourist entries in protected areas in the country (Heinen and Kattel, 1992).

4.4. The convention on the conservation of migratory species of wild animals

The Convention on the Conservation of Migratory Species of Wild Animals, also known as CMS or the Bonn Convention, came into force in 1979; as of 2009, there were 112 Parties (www.cms.int). Throughout its history, CMS has attracted fewer Parties than the other Conventions described here, in part because many nations of the Western Hemisphere were already party to an older regional convention protecting migratory wildlife (the Western Convention of 1940). As such, most Parties to CMS are in the Eastern Hemisphere, but more recently, a number of Latin American countries have ratified it. As the name implies, CMS focuses on migratory wildlife and not with protected areas *per se*. It is thus much more of a species-based than area-based conservation convention, but within its 20 Articles there are some clauses that are germane for the topic at hand.

CMS's Article 1 considers conservation status to be favorable if (among other things) the distribution and abundance of migratory species approach historic coverage, suitable ecosystems for conservation exist, and that there is sufficient protected habitat to maintain migratory species. The Article further describes unfavorable conservation status as being those in which these (above) conditions are not met. CMS's fundamental principles (Article 2) similarly contains an important clause outlining the importance of conserving habitat: The Parties acknowledge the importance of migratory species . . . "taking individually or in cooperation appropriate and necessary steps to conserve such species and their habitats."

Through its long history, and in conjunction with other Conventions (especially Ramsar) CMS has been indirectly importance in expanding protected area networks, and especially in Europe and Africa due to the large avian migrations between those two continents. A number of small reserves were created along flyways that offer staging and stepping-stone habitats, and many of these also contain significant wetland resources. None-the-less, with its focus on migratory species, its relatively few signatory nations, and its appendices of species under varying degrees of threat, it is not nearly as important for protected areas as the other instruments described here, but is important for species protection. Under its auspices, various important regional agreements have been established and form some rather interesting case studies in species (and area) conservation. Among these are: the Agreements for the Conservation of Cetaceans in the Black and Mediterranean Seas and Contiguous Atlantic Area; the Africa-Asia Migratory Waterbird Agreement, several agreements on sea turtles in the Pacific and Indian Oceans, and the Agreement on Gorillas and their Habitats. Some habitat protection clauses are found within all of these.

4.5. The United Nations Convention on Biological Diversity

The United Nations Convention on Biological Diversity (CBD), formulated prior to and during the 1992 United Nations Conference on Environment and Development in Rio de Janeiro, is far and away the broadest of the international conservation agreements. It came into force on 29 December 1993 and has three main objectives: to conserve biological diversity, to use biological diversity sustainably and to share the benefits of biological diversity equitably (www.cbd.in). The CBD currently (2009) has 156 Parties. Many provisions of the Convention do not deal with protected areas *per se*, so I only highlight important aspects of CBD and focus on those few aspects that do relate to this topic.

Of its 42 Articles and 3 Annexes, Article 8 (*In Situ* Conservation) is the main one dealing with protected areas. Therein, contracting Parties are encouraged, as far as possible and as appropriate, to, establish systems of protected natural areas and to develop guidelines for their selection and management.

Article 8 of CBD further requests Parties to manage important biological diversity appropriately whether it is located within the protected area network or not, and to promote general ecosystem protection. Parties are also requested to promote environmentally sound sustainable development in the vicinity of protected areas, to promote restoration of degraded ecosystems, and to control exotic species that pose a risk to conservation. Parties are further encouraged to use innovative practices in management and to involve local and indigenous communities in protected areas management. The final clauses of Article 8 instruct Parties to develop appropriate regulatory legislation to conserve endangered species, cooperate in financial support for *ex situ* conservation, and regulate processes that may adversely affect biological diversity in accordance with Article 7, which addresses the identification and monitoring components of biological diversity. Annex 1 (referenced in Article 7) defines components to be monitored to include ecosystems and habitats with high diversity, large numbers of endemic or threatened species, wilderness, and/or important habitat for migratory species. It further instructs Parties to identify and monitor communities and species that are threatened, contain wild relatives of important domesticates or other value, or are important for research, conservation and sustainable use.

Much of the rest of CBD deals with issues of domestic biodiversity, genetic complexes, appropriate uses of biological diversity and trade, *ex situ* conservation, equitability and sustainability management of biodiversity, and not with protected areas *per se* (Glowka *et al.*, 1994). None-the-less, CBD is far and away the broadest in scope of any conservation treaty and recognizes explicitly that protected natural areas are essential for biodiversity conservation at all levels of integration (*i.e.* from genes to ecosystems) and such protected area systems are important to effectuating the CBD objectives. It is also useful to note that CBD articulates quite well the more modern view of protected areas as places in which human are an integral part, as opposed to the older view of raw nature and 'fences and fines' management characteristic of the first American model. Shortly after CBD came into force, WCPA modernized its protected area categories (above). Another aspect of CBD that has proven important since its passage is that many Parties have undertaken the task of creating national conservation strategies and action plans, which is promoted by Article 7 and Annex 1. Such plans have allowed for a fuller inventory of important biological diversity, and have generally contained parts that deal with the existing protected areas network of each Party, with recommendations for expansion and for better management of existing units.

5. General discussion and overview

From its humble American beginnings in 1872, the international movement to conserve protected areas at the national level has mushroomed in the modern era. Most nations now maintain systems of protected areas, and the majority are now Parties to all of the international conservation conventions discussed above. The intellectual breadth of protected area management categories recognized worldwide, and of the types of both traditional and non-traditional uses permitted within them, has also expanded greatly, as has the use of zoning large areas to permit more or less uses in specific tracts depending both on the need for biological conservation and human enterprises. None-the-less, the stamp of the earlier history of a Western and largely American model still pervades many protected area systems. For example, of the categories recognized worldwide, four were derived largely from American law. These include Category Ib (Wilderness), II (National Parks); III (Natural Monuments) and to a lesser degree, Category IV (Wildlife Reserves, refuges, managed habitat areas, *etc.*). Categories V (Managed Landscapes and Seascapes) also had some precedent in the national protected area system of the United States, with its National Seashores, National Recreation Areas, *etc.* The nation that began the movement is thus still the most dominant player, at least in terms of general categories and many accepted management practices.

However, the internationalization of the protected areas movement created many opportunities and altered many previously accepted practices, which proved important for the continuous expansion of protected area systems. Allowing private in-holdings and some limited extractive uses from National Parks, for example, did not become recognized until the 1980s, and was only recognized by WCPA as a result of several national experiments to remove local people completely from Category II reserves. One such case happened in Nepal, where two local villages were removed from high-altitude Rara National Park, and their residents relocated into the western Terai (lowlands) of the country. Within a generation, over half of the original population had either left or died from malaria or other lowland diseases, and Nepal's Department of National Parks and Wildlife Conservation ended any plans to remove larger (but still rather small) local populations from other Himalayan national parks such as Langtang and Everest (Heinen & Kattel, 1992). The international movement, and the international organization in the form of IUCN's WCPA, was sensitive to these issues and adjusted Category II accordingly by allowing the zoning out of traditional villages (*i.e.* they are not recognized as part of the park, although they are surrounded by it). Similarly, the development of Category VI in 1994, and its subsequent worldwide expansion, was an important acknowledgement of the needs of many people in developing countries by recognizing that traditional local uses, and even some more modern commercial, uses of at least some protected areas should be permitted.

The literature on many facets of protected area management is similarly expanding greatly, and both space and topical content of this volume does not permit a closer look at some of the more scientific aspects. Suffice it is to say that the literature in conservation biology, a field only recognized since the 1980s, includes literally hundreds of well-done studies on issues such as placement of reserves, how to prioritize areas for protection based on scientific criteria, the appropriate size of reserves, the utility of maintaining natural corridors to promote gene flow between reserves, the placement and uses of buffer zones, *etc.* (Primack, 2006). Modern conservation agencies thus have at their disposal a great arsenal to help them plan the most efficient uses of scarce resources in conserving biodiversity. But conservation biology, first defined by Michael Soule (one of its founders) as a "mission oriented, crisis discipline" also recognizes that time is running out for many wild species on earth, and for the places that harbor them.

We are in the midst of a mass extinction, recognized by science as the 6th such event in the history of life on Earth, and the only one to be caused by one species: ourselves (Wilson, 1999). Modern humans threaten to have a commensurate impact, albeit more slowly, of the great asteroid that landed in the Western Caribbean and wiped out the dinosaurs - and about half of all other life forms - some 65 million years ago. While much conservation literature, and the Convention on Biological Diversity, also discusses the importance of *ex situ* conservation in the form of seed banks, zoos, botanical gardens, *etc.* (and doubtlessly they are all important), science and much of society recognizes that *in situ* conservation of species - and complexes where they occur naturally - is a much more cost effective and efficient way to conserve biodiversity. It has the added advantage of keeping ecological phenomena (*e.g.* predation, competition, migratory behavior, pollination) intact (or at least partly so) and allows for the evolutionary game to continue. *Ex situ* conservation provides, at best, a short term buffer.

So the international movement to conserve protected areas will increase over time. As the planet becomes increasingly crowded, it will remain the major way to conserve biodiversity and, ultimately, ourselves. New ideas - and ideals – are constantly expanding the field with more recent foci on issues such as landscape-level conservation across wide regions, a recent major increase in the study of invasive species and their removal, and better ways to manage areas already protected. But like the situation with so much else, the legal and policy instruments, both within nations and across nations in the form of international treaties and programs, lag greatly behind the science (*e.g.* Jacobson and Weiss, 2000). To further this field, we will need much more land set aside, which may mean including even broader protected area categories recognized in the future, and we will need to pay much more attention and legal recognition to dynamic processes across landscapes. In short, we need to become much smarter and more adept at living with the functional natural world.

Author details

Joel Heinen
Department of Earth and Environment, Florida International University, Miami, FL, USA

6. References

Alonzo, J. & J.T. Heinen. 2011. Miami-Dade County's Environmentally Endangered Lands Program: Local efforts for a global cause. Natural Areas Journal 31(2): 500-506.

Baral, N., M. Stern and J.T. Heinen. 2007. Integrated conservation and development project life cycles in the Annapurna Conservation Area, Nepal. Biodiversity and Conservation 16(10):2903-2917.

Batisse, M. 1982. The biosphere reserve concept: A tool for environmental conservation and management. Environmental Conservation 9:101-114.

Batisse, M. 1986. Developing and focusing the biosphere reserve concept. Nature and Resources 22(3):1-11.

Borgerhoff Mulder, M. & P. Coppolillo, P. 2006. Conservation: Linking Ecology, Economics and Culture. Princeton University Press, Princeton, NJ.

Bruntland, G.H. 1987. Our Common Future. Oxford University Press, Oxford, UK.

Bruyere, B. L. A. Beh & G. Lelngula. 2009. Differences in perceptions of communication, tourism benefits and management issues in a protected area in rural Kenya. Environmental Management 43(1): 49-59.

Campbell, L. M. & A. Vainio-Mattila. 2003. Participatory development and community-based conservation: Opportunities missed for lessons learned? Human Ecol. 31(3): 417-437.

Chape, S., M. Spalding & M. Jenkins. 2008. The World's Protected Areas: Status, Values and Prospects for the 21st Century. Berkeley: University of California Press.

De Boer, W. F. & D. S. Baquete. 1998. Natural resource use, crop damage and attitudes of rural people in the vicinity of Maputo Elephant Reserve, Mozambique. Environmental Conservation 25(3): 208-218.

Dombeck, M. P., C. A. Woods & J. E. Williams. 2003. From Conquest to Conservation: Our Public Lands Legacy. Island Press, Washington, DC.

Dorsey, K. 1998. The Dawn of Conservation Diplomacy: U.S. – Canada Wildlife Protection Treaties in the Progressive Era. University of Washington Press, Seattle, WA.

Faure, M. and J. Lefevere. 1999. Compliance with international environmental agreements. Pages 138-156 in Vig, N. J. and R. S. Axelrod, editors. The global environment: institutions, law, and policy. Congressional Quarterly, Inc., Washington, DC.

Fiallo, E. A. & S. K. Jacobson. 1995. Local communities and protected areas: Attitudes of rural residents towards conservation and Machalilla National Park, Ecuador. Environmental Conservation 22(3): 241-249

Fischman, R. L. 2003. The National Wildlife Refuges: Coordinating a Conservation System through Law. Island Press, Washington, DC.

Fu, B. J., K. L. Wang, Y. H. Lu, K. M. Ma, L. D. Chen & G. H. Liu. 2004. Entangling the complexity of protected area management: The case of Wolong Biosphere Reserve, southwestern China. Environmental Management 33(6): 788-798.

Gillingham, S. & P. C. Lee. 2003. People and protected areas: A study of local perceptions of wildlife crop damage conflict in an area bordering the Selous Game Reserve, Tanzania. Oryx 37(3): 316-325.

Glowka, L., F. Burhenne-Guilmin, and Synge, H. 1994. A Guide to the Convention on Biological Diversity. IUCN Publications, Gland, Switzerland.

Groom, R. & S. Harris. 2008. Conservation on community lands: The importance of equitable revenue sharing. Environmental Conservation 35(3): 242-251.

Hails, A.J. 1996. Wetlands, Biodiversity and the Ramsar Convention: The Role of the Convention on Wetlands in the Wise Use and Conservation of Biodiversity. IOUCN Publications, Gland, Switzterland.

Heinen, J.T. 1996. Human behavior, incentives and protected area management. Conservation Biology 10(2): 681-684.

Heinen, J.T. & B. Kattel. 1992. Parks, people and conservation: AS review of management issues in Nepal's protected areas. Population and Environment 14(1): 49-84.

Heinen, J.T. & J. N. Mehta. 1999. Conceptual and legal issues in the designation and management of conservation areas in Nepal. Environmental Conservation 26(1): 21-29.

Heinen J.T. & S.K. Shrestha. 2006. Evolving policies for conservation: an historical profile of the protected area system of Nepal. Journal of Environmental Planning and Management 2006; 49(1):41-58

Heinen, J. T. and R. Vande Kopple. 2003. Profile of a biosphere reserve: The University of Michigan Biological Station and its conformity to the Man and Biosphere Program. Natural Areas Journal 23(2):165-173.

IUCN. 2003. United Nations List of Protected Areas. Gland, Switzerland, IUCN Publications.

Jacobson, H. K. and E. B. Weiss. 2000. A framework for analysis. Pages 1-18 in Weiss, E. B. and H. K. Jacobson, editors. Engaging countries: strengthening compliance with international environmental accords. MIT Press, Cambridge, MA.

Klemm, C. de and C. Shine. 1993. Biological diversity: legal mechanisms for conserving species and ecosystems. IUCN - The World Conservation Union, Environmental Law and Policy Paper No. 29, Cambridge, UK.

Lepp, A. 2007. Residents' attitudes towards tourism in Bigodoi village, Uganda. Tourism Management 28(3): 876-885.

Lepp, A. & S. Holland. 2006. A comparison of attitudes toward state-led conservation and community-based conservation in the village of Bigodi, Uganda. Society & Natural Resources 19(7): 609-623.

McShane, T.O. & M.P. Wells. 2004. Getting Biodiversity Projects to Work: Towards more Effective Conservation and Development. Columbia University Press, New York.

Norton, B.G. 2005. Sustainability: A Philosophy of Adaptive Ecosystem Management. University of Chicago Press, Chicago, IL.

Primack, R. B. 2006. Essentials of Conservation Biology (4th edition). Sinaur: Sunderland, Mass.

Redford, K. H. & S. E. Sanderson. 2000. Extracting humans from nature. Conservation Biology 14(5): 1362-1364.

Sherman, P. B. & J. A. Dixon. 1990. Economics of Protected Areas: A New Look at Costs and Benefits. Washington, DC: Island Press.

Sitarz, H.I. 1993. Agenda 21: The Earth Summit Strategy to Save our Planet. EarthPress, Boulder, CO.

Susskind, L. E. 1994. Environmental diplomacy: Negotiating more effective global agreements. Oxford University Press, Oxford, UK.

Swanson, T. 1997. Global action for biodiversity. IUCN Publications, Cambridge, UK.

Wilson, E.O. 1999. The Diversity of Life (New Edition). Harvard University Press, Camridge, MA.

Wittemyer, G., P. Elsen, W. T. Bean, A. Coleman, O. Burton & J. S. Brashares. 2008. Accelerated human population growth at protected area edges. Science 321(5885): 123-126.

Zaslowsky, D. & T.H. Watkins. 1986. These American Lands: Parks, Wilderness and Public Lands. The Wilderness Society, Washington, DC.

Chapter 2

New Issues on Protected Area Management

David Rodríguez-Rodríguez

Additional information is available at the end of the chapter

http://dx.doi.org/10.5772/48269

1. Introduction

Protected areas (PAs) are the main strategy to face the global deterioration of ecosystem goods and services due to human causes (Mora & Sale, 2011; Sanderson et al., 2002) and especially of biodiversity (Jenkins & Joppa, 2009; Pullin, 2002).

Almost 150 years have passed since the designation of the first "modern" PAs, back in the XIXth century. Designation of PAs has growth rapidly since that date and especially since the last third of the XXth century (Jenkins & Joppa, 2009; McDonald & Boucher, 2011). Currently, there are over 120,000 nationally and internationally-designated PAs in virtually all the countries of the world (UNEP-WCMC, 2008). They covered about 31.235.000 km^2, over 21% of the global terrestrial (approximately 17,290,000 km^2) and ice-covered area (some 13,950,000 km^2) by 2010, according to the Wold Database on Protected Areas (WDPA), and continue to expand. Conservation has become one of the main land uses globally, with approximately 13% of land under some form of protection (Jenkins & Joppa, 2009; McDonald & Boucher, 2011). Vast human, material and economic resources are allocated to conserve and use sustainably the natural and cultural resources provided by PAs. And yet biodiversity and the other ecosystem goods and services continue to be lost (Butchart et al., 2010). As a result, PA management is receiving increasing attention as one of the key aspects for the effective conservation of PAs.

In this chapter, different aspects related to PA management are discussed. It also tries to clarify some controversial concepts and to encourage the discussion on a number of challenging issues of interest for scientists, managers, policy-makers and conservationists.

2. Definition of protected areas

There are two internationally accepted definitions of PA: the definition given by the Convention on Biological Diversity (CBD, 1992), and the latest definition proposed by the IUCN (Dudley, 2008). Both have subtle conceptual differences that should be scrutinized in detail to know what exactly is being talked about:

a. According to the CBD, a protected area is "a geographically defined area which is designated or regulated and managed to achieve specific conservation objectives" (CBD, 1992). This broad definition could apply to virtually every space with some regulation to achieve proper use and conservation of resources, such as game reserves, managed logging areas, marine reserves, or Biosphere Reserves, for instance. No specific mention to the conservation of biological features is given. As a result, this definition could also apply to non-biological resources meriting conservation, such as physical, geo-morphological or even cultural features, like museums or cathedrals.
b. According to the most updated definition of PAs by the IUCN, a PA is "is a clearly defined geographical space, recognised, dedicated and managed, through legal or other effective means, to achieve the long term conservation of nature with associated ecosystem services and cultural values" (Dudley, 2008). This definition is more precise, and includes three key and advanced conservation concepts: 1) the "long-term" scope of conservation; 2) the specific conservation of "nature"; and 3) the conservation of complementary ecosystem and cultural goods and services.

All six (seven?) PA management categories proposed by the IUCN (Dudley, 2008) share the principles included in this definition. This second definition of PA is more accurate and thus it will be used along this chapter.

3. Multiple use. Sustainable development in protected areas?

Currently, almost no one argues that the overarching goal of PAs is conserving biodiversity and that any other management objective should be subject to biodiversity maintenance, improvement or restoration (Múgica & Gómez-Limón, 2002). There is, however, considerable debate on the weight of additional social and economic objectives in PA management, following the desirable yet vague concept of sustainable development (WCED, 1987).

Global agreements are giving PAs a starring role in many aspects other than biodiversity conservation. One extended mission included in its most updated definition is providing while preserving ecosystem goods and services and associated cultural features other than biodiversity. The most relevant goods and services provided by PAs include: raw materials, food, genetic, medicinal and ornamental resources, water purification, air quality regulation, erosion prevention, mitigation of extreme events, pollination, biological control, carbon sequestration, soil formation, primary production, and nutrient cycling (Chape et al., 2008; Millenium Ecosystem Assessment [MEA], 2005; Naughton-Treves et al., 2005). Lately, PAs are also increasingly conceived as optimum testing fields for monitoring global change (Alcaraz-Segura et al., 2009), as well as useful and cost-effective means for climate change adaptation and mitigation (Dudley et al., 2010). As seen, the "environmental targets" of PAs are quite ambitious. Not all PAs are able to provide all those goods and services, although some of them, especially the biggest ones, might.

Nevertheless, following the dominant tri- (or four-) dimensional concept of sustainability (Spangenberg, 2002), PAs are also entrusted a high number of other social and economic roles, multiplying management objectives virtually to the infinite. Some of those roles are: scientific research, recreation and tourism, inspiration for art and culture, spiritual and cognitive development, aesthetic enjoyment, preservation of traditional cultures and practices, improving social welfare, enhancing environmental education and awareness, promoting peace and security, facilitating people's participation and governance, and boosting economic development at multiple scales (Chape et al., 2008; Naughton-Treves et al., 2005).

Whereas attaining all those targets would be ideal for any PA, the challenge for PAs to do so is overwhelming. These goals are so wide and ambitious that, even in the cases where they are not directly opposed (UNESCO, 2002), it looks doubtful that any piece of land in the world could comply harmoniously with all of them.

Acknowledging the role of man within nature and its importance for the conservation of many species, habitats and ecological processes, especially in Europe where cultural landscapes are paramount (Jongman, 2002), multiple use of PAs sometimes becomes a "dogma" for PA planners, managers and policy-makers, irrespective of very different conditions and situations. Most conservation policies try to force sustainable development to happen, often through the zoning of PAs (Naughton-Treves *et al.*, 2005; UNESCO, 2002) or the placing of tourist infrastructures (Farrell & Marion, 2001), regardless of the fact that some activities may conflict with conservational ends. Such conflicts are frequent in densely populated areas between conservation and recreation activities (Tisdell, 2001).

Figure 1. Cultural features and traditions are an important and often neglected asset of protected areas

Cultural, recreational, educational and economic values should be pursued in PA planning and management, as long as the key aim of PAs (*i.e.*, conserving biodiversity) is never left behind to satisfy other possibly conflicting objectives. If social and economic objectives are compatible with biodiversity conservation or improvement, we should try to integrate those within planning and management. If they are not, their practice attainment should be considered no more than theoretical goodwill, no matter how temping development promises may look like. If poorly conceived, executed and regulated, even the best socioeconomic proposal might turn out to be the deadliest ecological error (Witte et al., 1991).

Therefore, predicted future trends on the prevailing designation of "multiple use" PAs (IUCN categories V and VI) in the years to come (McDonald & Boucher, 2011), although a good piece of news in itself, might entail subtle risks for effective conservation of resources.

The best solution to make most uses of PAs compatible would be designing a territorial model made of different types (regarding their priority function) of large, zoned PAs connected through biodiversity-oriented managed landscapes (Mata et al., 2009; Rodríguez-Rodríguez, 2012b). However, limited resources and high land and visitor pressure, especially in densely populated areas, force us to put the preservation of ecosystem processes first and before any other use or consideration of PAs (Pressey *et al.*, 2007).

4. The role of tourism in protected areas

Tourism is an extremely delicate issue regarding PA management. Whereas often seen and promoted as social and economic salvation for local communities, if unregulated it may, on the one hand, lead to the deterioration or destruction of the resources of the PA and, on the other, threaten local culture. If well planned and managed, however, tourism can provide a significant source of revenue for local populations and / or PA administrators, as well as increase visitor education and environmental awareness (Chape et al., 2008).

Figure 2. Visitors may be an intense pressure in protected areas

Nevertheless, management and conservation problems are intrinsic to tourist visitation of PAs. These include: increasing use of resources, including land-use changes for developing tourism infrastructures in or around PAs; soil trampling and compacting; vegetation removal; animal disturbance; soil erosion; littering; noise making; introduction of alien species and varieties; damage to geological features, cultural sites, vegetation or public use infrastructure; increased fire risk; air pollution from transportation; animal road killing; changes in wildlife behavioural patterns due to human habituation, among others impacts (Chape et al., 2008; Farrell, 2002). Some of these impacts can be so serious that they can compromise the long-term conservation of the resources being protected (Barrado, 1999; Rodríguez-Rodríguez, 2009; Tisdell, 2001). Visitor impacts may also result in a reduced quality of recreational experiences and in conflicts among visitors (Farrell 2002; Phillips, 2000).

Additionally, public use management competes for scarce funding with other management and conservation objectives of PAs, as it is a highly resource-demanding activity (VVAA, 2000).

While zoning PAs and building tourist infrastructure might be useful at reducing impacts from tourism in PAs by directing main fluxes towards the least fragile zones of the PA (Farrell, 2002; UNESCO, 2002), they cannot usually prevent invasion of more vulnerable zones (Barrado, 1999). In fact, tourism infrastructure can attract even more visitors to the PA, and may result in its overcrowding and collapse (Morales, 2001).

As a result, although public use in PAs is a desirable and broadly-accepted concept, it should be strictly regulated or even prohibited in the cases when it is incompatible with conservation targets which must prevail in PAs.

5. Free access to protected areas? Financial and equity aspects of entrance fees to protected areas

As we just saw[1], impacts of visitation to PAs are numerous and serious (Chape et al., 2008; Rodríguez-Rodríguez, 2009). Whereas there is free access to most PAs in industrialised countries, no matter what land ownership is, entrance fees are common practise in other parts of the world (Emerton et al., 2006).

There are a number of reasons in favour of charging a fee for accessing PAs. First and most important, entrance fees are effective visitor filters (Emerton et al., 2006). On the one hand, they enhance the quality of visitors by discouraging the least interested visitors often having the poorest environmental awareness (Barrado, 1999) from entering the PA, thus helping prevent most unsocial behaviour (McKercher & Weber, 2008). On the other, they potentially reduce the quantity of visitors, thus diminishing the human pressure on species, soil, vegetation, ecosystems, cultural features and infrastructure (Rodríguez-Rodríguez, 2009).

The second reason, as seen at the beginning of this chapter, is that PAs provide multiple social goods and services related to improved health and well-being: recreation, tourism, sport, relax, inspiration, cultural and aesthetic enjoyment, and spiritual and cognitive development (Chape et al., 2008). The fact that most of these goods and services have not

---

[1] See heading 4

market price does not mean their economic value is negligible; rather on the contrary (Emerton et al., 2006; Kerry et al., 2003). Thus, charging an amount of money to enjoy all these benefits from PAs does not seem exaggerate.

Additionally, paying an amount of money makes the visitor conscious of the value of the visited place (Emerton et al., 2006). It gives him also a temporal "membership" to a "selected" group of people (club) with the right to respectfully enjoy the wonders of the PA.

Finally, most PAs are underfunded (Leverington et al., 2010; Nolte et al., 2010) and budgetary restrictions will tend to be the rule more than the exception in the future due to financial constraints in the public sector which currently makes up most PA funding globally (Phillips, 2000). Thus, provided that incomes from visitors are properly invested in PAs (in the same PA they are expended or in other PAs to support underfinanced or undervisited PAs), PA administrators could count on an additional source of financing to enhance management and conservation and to provide new employment opportunities (Emerton et al., 2006; Phillips, 2000).

Fee collection should be carefully planned and implemented. It must consider adequate equity prior to its implementation so none will be deprived of his or her right to enjoy PAs. Thus, discounts should be applied to low-income visitors upon appropriate certification. Dual-pricing policies charging foreign visitors higher than local ones are an equitable and socially acceptable option in developing countries (Walpole et al., 2001).

Well-thought decisions should also be taken about visitors being charged repeatedly during their visits (for instance, for car parking, access to the PA, and / or visitation of public use infrastructure), and about the appropriate quantities to be charged in view of recovery costs, so no additional charges are imposed to the PA administration (Chape et al., 2008).

Figure 3. Parking fee in Peñalara Natural Park, Madrid Region, Spain

6. Effective management vs. effective conservation

There is not consensus on the effectiveness of PAs as a global strategy for biodiversity conservation. There are examples of conservation success and failure, although the latest are rarely reported (Mora & Sale, 2011). Lack of or inappropriate monitoring and assessment activities are common management deficiencies (Leverington et al., 2010) making the assessment of PAs often difficult, arbitrary or meaningless (Parrish et al., 2003).

Therefore, whereas the protection of spaces for biodiversity conservation remains a valid and useful strategy to mitigate current environmental crisis, a general statement in favour of the actual effectiveness of PAs as a tool for biodiversity conservation is simplistic. Rather, PA effectiveness should be analysed case by case[2].

It is generally accepted that legal designation plus effective management result in the effective conservation of PAs (Hockings et al., 2006). It is not always the case, however. There are important pressures and threats to the conservation of PAs and their resources that operate at a scale broader than PAs and that may, therefore, spoil brilliant management efforts, as they are outside the scope, means and competence of PA managers (Alcaraz-Segura et al., 2009; Jameson et al., 2002; Parrish et al., 2003).

Impacts of global change are posing additional pressures on the conservation of biodiversity worldwide (Araújo et al., 2011). Regional impacts can also cause severe effects on faraway zones regardless of their degree of protection or management effectiveness. Acid rain on land and oil spills on sea are good examples. As a result, a PA can be legally designated, well-planned and efficiently managed and see its resources being degraded (Mora & Sale, 2011), more so in the marine environment where connectivity is higher than on land (Jameson et al., 2002).

PAs exchange matter, energy and information with their surroundings. These exchanges are vital to ecological processes underpinning biodiversity and other ecosystem goods and services (Múgica et al., 2002). Shifts in nature, extent, direction and intensity of these interactions may also, however, result in significant alteration of the ecosystem structure and function, even if optimum PA management is in place. Therefore, PAs cannot be managed in isolation from the surroundings influencing those (Radeloff et al., 2010). As a result, nearly as much care should be taken when managing zones surrounding PAs as when managing PAs themselves.

Effective management can play an important role in PA conservation. It can also contribute to adaptation and mitigation of some regional or global pressures and threats, such as climate change, through PAs (Dudley et al., 2010). However, the role of management in attaining long-term conservation of resources should not be overemphasized. PA managers can very rarely achieve effective conservation of managed sites on their own. In contrast, the importance of achieving sustainable development at all scales, integrating PAs in a wise and wider territorial planning (Rodríguez-Rodríguez, 2012b) cannot be stressed enough.

---

[2] See heading 7

7. Assessing protected areas. Scales and perspectives

For too many years, it was assumed that the resources harboured by PAs were effectively conserved just because they were legally protected. Regrettably, empirical and scientific evidences refute the previous assumption (Butchart et al., 2010). Luckily, that vision is currently outdated and much literature has been written on "paper parks", although they still prevail in certain contexts (Bonham et al., 2008; Davis, 2001).

Currently, the question is: what should be assessed? Or: how to measure PA effectiveness?

Multivariate, socio-ecological systems such as PAs are difficult, costly and time-consuming to characterize, monitor and assess. Lack of basic knowledge increases the uncertainty of what to monitor, how and why (Parrish et al., 2003).

As a result, developing a complete, ecologically-sound and meaningful-for-management assessment system has proved extremely complicated, despite several efforts developed worldwide: Hockings et al. (2000; 2006); Ervin (2003a), Parrish et al. (2003), Pomeroy et al. (2005), Gaston et al. (2006); Mallarach et al. (2008); Ioja et al. (2010); Rodríguez-Rodríguez & Martínez-Vega (in press).

To start untangling this issue, a distinction between different concepts which are often used indiscriminately (Ervin, 2003b) should be made. One is "management effectiveness" and the other is "PA effectiveness". Identifying both concepts would be the same as saying that all a PA can do to protect and conserve its resources is due to management. Once more, evidence does not seem to support this statement (Jameson et al., 2002; Mora & Sale, 2011), as we just saw[3].

There are unprotected places well conserved due to lack of human visitation, accessibility and exploitation (Sanderson et al., 2002). There are paper parks enjoying a good conservation status for similar reasons. And there are legally-designated, well-managed PAs being degraded due to unsustainable socioeconomic contexts and / or regional or global pressures and threats (Jameson et al., 2002; Radeloff et al., 2010).

Thus, PA effectiveness is not the same as management effectiveness. Stating so would put all the responsibility for the conservation status of PAs on managers with limited means, capacity and competences to cope with many factors outside their control. Thus, on the one hand, this identification is not precise and, on the other, it may neither be fair to PA managers (Rodríguez-Rodríguez & Martínez-Vega, in press).

Additionally, according to the "Management Effectiveness Evaluation Framework" (Hockings et al., 2000, 2006), protected area management effectiveness evaluation (PAME) should be target-driven. Therefore, if considered precisely, PAME could not be applied to most PAs worldwide lacking specific, clear, mensurable management objectives (Bertzky & Stoll-Kleemann, 2009; Pomeroy et al., 2005).

---

[3] See heading 6

Therefore, "management effectiveness", which is an important part of "PA effectiveness", should only deal with factors directly linked to management which managers can control and address with sufficient knowledge, capacity and resources, while leaving out context variables they cannot manage or even influence.

There are different conceptions of "PA effectiveness": from its identification with "management effectiveness" (Hockings et al., 2000; Ervin, 2003a; Hockings et al., 2006), to "ecological effectiveness" (Gaston et al., 2006), "ecological integrity" (Parrish et al., 2003), or "sustainability"[4] (Rodríguez-Rodríguez & Martínez-Vega, in press).

Language precision is fundamental in science, but so far none has come up with an ideal, agreed term to designate overall "PA effectiveness" despite the wide literature on the topic: are we measuring effectiveness?, performance?, functionality?, conservation potential?, sustainability?, success?, ecological integrity?, accomplishment? This uncertainty is because the core object being assessed is complex and integrates several different parameters (often environmental and also social and economic) not easily defined in one known-term.

A second question arises that further highlights the complexity of assessing PAs in an integrated and scientific manner: is "effectiveness" a common, "objective" concept? Is it site-specific and thus, "subjective"? Or can it be both? The answer is not straightforward. There are different perspectives on PA effectiveness assessment closely linked to the scales considered and the objectives pursued with the assessment.

Scales are important when defining PA effectiveness. For example, "management effectiveness", "ecological effectiveness" and "ecological integrity" are usually conceived as "site-specific" due to the different ecological conditions, contexts and, therefore, management needs and objectives of every PA (Hockings et al., 2006). In contrast, "sustainability assessment" proposes a common assessment and valuation method for PAs belonging to the same context (administrative, bio-geographical, or socioeconomic). Based on the most updated definition of PA (Dudley, 2008), it assesses the likelihood that a PA can conserve its natural and associated cultural resources and ecosystem services in the long-term on relevant common parameters for management and decision-making. It is also a site assessment, but it allows comparison and prioritization among the individual PAs assessed as assessment is made upon the same parameters (Rodríguez-Rodríguez & Martínez-Vega, in press).

At a broader scale, the effectiveness of conservation networks should also be assessed using landscape metrics and indicators (Burel & Baudry, 2003).

---

[4] In our preliminary work (Rodríguez-Rodríguez & Martínez-Vega, in press), we identified "environmental" or "hard" "sustainability" as the core concept to be assessed when evaluating PAs, according to the goal of long-term conservation of natural and cultural resources established in the latest definition of PAs by the IUCN (Dudley, 2008) and in the original definition of sustainable development (WCED, 1987). However, some colleagues were reluctant to the use and formalization of the term "sustainability" as a result of the vagueness of its original definition and its different interpretations, so we changed the term to "PA effectiveness", also meaning the PA capacity or potential to conserve its resources in the long-term. Nevertheless, we think "sustainability" is a valid focus for assessing PAs with no more difficulties in its definition and formalization than any other complex integrated concept such as "effectiveness".

Table 1 summarizes the different complementary approaches to assessing PA effectiveness.

Assessment	**Scale**	**Object**	***Main* target group**
Ecological effectiveness	Individual PA (Site-specific)	Ecological parameters: ecosystems, species and the physical-chemical environment	*Scientists*; PA managers
Ecological Integrity	Individual PA (Site-specific)	Threats; Focal species, communities and ecological systems	*PA managers*; decision-makers
Management effectiveness	Individual PA (Site-specific; Comparable)	Management goals, objectives and strategies regarding: context, planning, inputs, processes, outputs & outcomes	*PA managers*; PA network managers; decision-makers
Sustainability / Effectiveness	Individual PA (Comparable)	State of conservation, planning, management, social and economic context, social perception and valuation, and threats to PAs	*PA network managers*; PA managers; decision-makers
PA network	Landscape; territory	Connectivity, biodiversity representativeness, gap analysis, etc.	*PA network managers;* decision-makers; territorial planners

Table 1. Complementary approaches to assessing protected area effectiveness

8. Protected area objectives vis à vis management objectives

Clear objectives are fundamental for adaptive management and, therefore, for effective PA conservation (Hockings et al., 2006; Pomeroy et al., 2005). Nevertheless, it is rare that clearly-stated, science-based objectives be specified in any PA norm or management document (Bertzky & Stoll-Kleemann, 2009; Pomeroy et al., 2005) and, when management objectives are set up, they are often vague, inadequate or even contradictory (Naughton-Treves et al., 2005).

Thus, the establishment of clearly-defined objectives at different levels in the long, medium and short-terms remains one the most urgent needs for effective management and conservation of PAs.

PA objectives should be initially specified in the designation norm of each PA, according to its conservation and management characteristics. Although these objectives might change in the medium-term due to the evolution of natural systems, they are mostly stable throughout the time, as they portrait a long-term vision of the PA. They can be similar to or adapted from the general objectives of the different PA management categories proposed by the IUCN (Dudley, 2008).

In contrast with these long-term "stable" objectives, medium-term specific objectives should be developed for each management planning period; *i.e.*, the validity period of an average management plan of 4-10 years (Thomas & Middleton, 2003). These medium-term planning objectives should try to make the long-term objectives of the PA happen considering the different circumstances of the PA throughout the time.

Finally, management plans are often disaggregated into more operational annual work-plans, whereby the medium-term objectives are made specific to the management means and to the reality of the PA through achievable and measurable short-term objectives (Thomas & Middleton, 2003). These short-term management objectives should also foresee (regarding means and probabilities of occurrence) and address sudden events not previously planned against which rapid action is required (Chape et al., 2008).

Figure 1 shows the conceptualization of the different types of objectives in PAs and their main characteristics.

Sensible objective definition leads to adequate planning and, if adequate means are available, to proper management of the PA. However, clear definition and precise forms of measurement of the different objectives are needed at all levels for management planning to be fully effective (Thomas & Middleton, 2003).

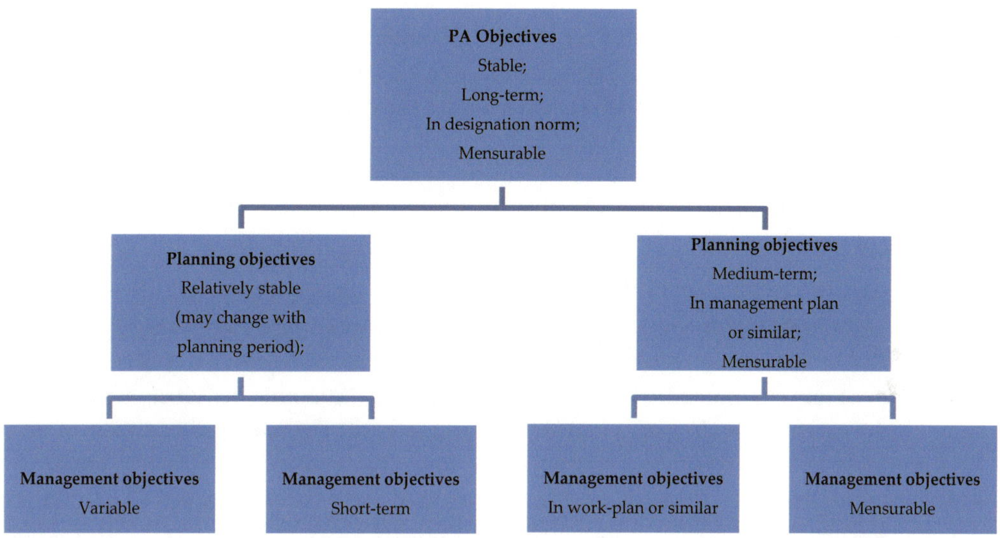

Figure 4. Outline of the different levels and objectives of PAs and their main characteristics.

9. The role of society in protected area management

PAs are portions of territory where multiple interests coexist and often collide (McKercher & Weber, 2010; Phillips, 2000). Although their main aim is to conserve biodiversity and the other ecosystem goods and services and associated cultural features, they include other environmental, social, economic and governance factors deserving attention. As such, they cannot and should not be managed in a unidirectional manner.

Neither decision-makers nor PA managers can cope with current challenges to PAs alone. An increasingly complex world requires increasingly innovative solutions, also regarding PA management. Thus, they should look for allies in other stakeholders if effective management and affordable enforcement is to be achieved.

Bottom-up, participatory approaches to PA management may be more resource and time-consuming than technical top-down approaches, but they are generally accepted better by all stakeholders and, as a result, they are often more effective and enduring. Consensus results in decisions being more legitimate, and in fewer conflicts (Fraser et al., 2006).

There are two priority groups whose close collaboration should actively be sought: scientists and local populations. Management measures can be more efficient, cost-effective and easily-implemented if they are based on sound science (Chape et al., 2008). Therefore, a stronger, deeper and more trustful relationship between PA managers and scientists is the first priority to enhance effective, participatory management (Zamora, 2010).

Local populations are also key to effective management (Rodríguez-Rodríguez, 2012a). Their views and attitudes towards PAs and their management can determine the success of most management initiatives (Múgica & Gómez-Limón, 2002; Borrini-Feyerabend et al., 2004).

Building a positive attitude towards PAs by local populations consists not only in informing individuals or associations (NGOs, neighbours, etc.) on management issues, but also in consulting them, making joint decisions and seeking their input to actually carry out some management activities (Múgica & Gómez-Limón, 2002). Environmental volunteering can be a very adequate way of participation of local populations into management, and also an effective means of reinforcing the identity of residents with the values, resources and governance of the PA.

Other relevant stakeholders to be considered in PA management are businessmen, politicians and tourists. New opportunities for business and employment related to the PA should be sought, mainly among local enterprises, as long as no negative impacts on the PA or its resources are guaranteed (Phillips, 2000). Therefore, economic activities affecting the PA should be carefully evaluated, planned and regulated in the management plan of each PA. Synergies among business, conservation and management practices should be potentiated; for instance, grazing in grasslands may have better ecological effects on biodiversity when complementing mowing than mowing alone (Metera et al., 2010). Thus, farmers can get free fodder for their animals and make profit by selling farm products associated with the PA, whereas PA managers can save money in ecosystem management.

Politicians should also be incorporated to valuating and managing PAs more actively. Their decisions can determine the future of a PA, including permission for aggressive activities within PAs or their surroundings, diminution or even disappearance of the protection status of a PA. Thus, they should be made aware on the importance of PAs and of their decisions for biodiversity and humans in the long-term. Social groups such as managers, scientists, local populations and NGOs can play an important role in lobbying for the cause of PAs.

Similarly, tourists should be made fully aware of their potential impacts on visited PAs (Hillery et al., 2001) through environmental education and volunteering, and public use infrastructure. Their collaboration with managers should be sought to obtain proper behaviour from them and, where possible, funding for the PA (Rodríguez-Rodríguez, 2012a).

10. Keeping pace with science

PA knowledge and management do not run at the same speed (Nolte et al., 2010; Pullin, 2002). Communication and trust between scientists and managers can also be much improved (Zamora, 2010). As a result, it is rare that sound, updated science be incorporated by managers and decision-makers and applied in everyday management (Pressey et al., 2007). Therefore, common management practices are frequently based just on experience, habit, or at best, on partial scientific knowledge (Pullin, 2002).

Sometimes, lack of scientific knowledge is due to overwork by managers. Too much field and / or administrative work leave them with little or no time or will to get updated on new management issues regularly (Pullin, 2002). Lack of adequate funding, interest or training programmes can also erode appropriate scientific knowledge among managers. Thus, staff training is an overall need in many countries (Leverington et al., 2010; Nolte et al., 2010).

And yet widespread updated scientific knowledge is vital for management and conservation effectiveness (Pressey et al., 2007). Resources devoted to conservation could be more successfully and wisely used if management practices were backed up by sound science (Pullin, 2002). Staff training should be considered an integral part of the management cycle (Kopylova & Danilina, 2011).

However, knowledge-transfer should not be unidirectional, from scientists to managers. The publication or communication of empirical management result and practices, be they successful or a failure, is also very important for advancing towards a more effective management of PAs and can provide science with interesting new research topics and experiences.

Figure 5. Presentations and workshops are useful training options

In conclusion, regular collaboration among scientists, managers and decision-makers is key for effective and efficient management (Pullin, 2002; Rodríguez-Rodríguez & Martínez-Vega, in press). Current disconnection and suspicion should be overcome if more effective ways of conserving PA are to be sought. This collaboration can take the form of regular working-groups or workshops. Ideally, researchers should be incorporated to the regular PA managing staff (Chape et al., 2008), be them restricted to specific PAs, covering a network of PAs, or both. Thus, actual management activities should be done considering both, sound science and best practice.

11. Rethinking protection scales: from protected areas to protected territories

Current *in situ* strategies for biodiversity conservation are not having enough success to stop or reverse biodiversity loss (Butchart et al., 2010; Mora & Sale, 2011), and thus biodiversity and the other ecosystem goods and services continue to be degraded (MEA, 2005).

Most PAs, especially in Europe, are too small to maintain viable populations within their boundaries in the long-term (Pullin, 2002). Under these circumstances, connectivity among PAs becomes fundamental (Jongman, 2002).

Ecological networks are a conceptual and practical step forward with regard to the initial concept of PA as islands in the territory. However, ecological networks are not without theoretical and practical shortcomings (Boitani et al., 2007). In addition, the current drivers causing environmental degradation are so intense than even ecological networks have been overcome by reality.

Enhanced connectivity of PAs through ecological corridors is thought to have increased PA conservation effectiveness (Beier & Noss, 1998). However, the wide range of scales taking part in ecological processes underpinning ecosystem goods and services (Pressey et al., 2007) make ecological corridors only partially effective, and only under some circumstances (Brunckhorst, 2000; Tzoulas et al., 2007). Moreover, the zones surrounding PAs are becoming increasingly artificial as a result of human activities and are therefore not accomplishing properly their buffer target (Radeloff et al, 2010). As a result, the effective protection and wise management of whole landscapes/seascapes surrounding PAs can be as important as the protection and management of PAs themselves (Jameson et al., 2002).

There is a growing consensus that effective biodiversity conservation cannot take place if complementary conservation scales broader than PA scale are not considered (Mata et al., 2009; Mora & Sale, 2011; Múgica et al., 2002; Rodríguez-Rodríguez, 2012b). Going from connected islands to connected landscapes is a desirable first step.

National conservation policies must be integrated and coordinated in order to be fully effective. The protection and management of isolated PAs, ecological networks or whole landscapes/seascapes alone is not sufficient for long-term conservation. Additional public conservation policies, mainly related to territorial planning, ecological restoration and environmental impact assessment (of projects, policies, plans and programmes), should complement the establishment and maintenance of PAs at the national level.

Full consideration of ecological processes should also be carefully considered and integrated in other public policies seriously affecting the steady provision of ecosystem goods and services such as housing, mining, fishing, transport or industry. Ecological considerations should always guide (and limit, when necessary) social and economic development policies and strategies if sustainable development is to be achieved (Naredo & Frías, 2005).

However, no matter how effective a national conservation policy is, its effectiveness will always be limited by regional or global trends, making local or national efforts only partially successful (Mora & Sale, 2011).

Future conservation strategies should be regional and global, as the driving forces pressing biodiversity are. Ecologically, it makes little sense planning and managing PAs restricted to local or national administrative boundaries. Thus, deeper and closer international collaboration is needed when planning for conservation.

Focus should change from PAs to protected territories of high conservation value following science-based criteria: species richness, diversity, taxonomic uniqueness, degree of threat, endemism, unusual ecological or evolutionary phenomena, or notable provision of ecosystem goods and services (Pullin, 2002). The ecorregional approach proposed by Olson et al. (2001) seems a promising and adequate scale to plan and manage conservation from an ecological perspective. This is the conservation approach followed by relevant conservation organizations, such as the Nature Conservancy (TNC, 2001).

Management should be much more collaborative and imply different stakeholders from different sovereign authorities. This form of closer international cooperation can have positive impacts in other public policies such as human development or defence (Sandwith et al., 2001).

Even if conservation and management strategies become global and effective, they will only be a limited solution to current and increasingly unsustainable human development trends (Millenium Ecosystem Assessment [MEA], 2005; Sanderson et al., 2002). Thus, if an enduring solution to the environmental crisis (and thus, sustainable development) is to be achieved, global determined and courageous measures must be taken in order to counter current population, development and consumption trends (Mora & Sale, 2011). All public policies must take into account the bio-physical limits of the Biosphere and they should be planned and implemented accordingly, considering the long-term consequences of today's decisions.

12. Final remarks: managing the Biosphere for conservation

The astonishing variety of life on Earth and other vital ecosystem goods and services are severely threatened by human activities well beyond the capacity and competence of PA managers, PA network managers or even decision-makers.

PAs are just part of the solution to the environmental crisis. We should not be excessively optimistic about their ultimate outcome as the main biodiversity conservation strategy. They are a shock therapy for a patient who is turning rapidly to worst.

In situ conservation strategies failed by protecting only some areas on diverse criteria, and leaving the rest of the territory without any form of development control or regulation. We should change (have changed, long ago) our way of thinking and practicing conservation. Conservation should be global, literally. I mean that the entire Biosphere should be protected and, where necessary, appropriately managed, leaving only a number of places of extraordinary social and economic value out of protection for intensive development and human activities, such as mines, cities, transportation corridors, some croplands and plantations.

Some may think this is asking too much, but all signs point to the fact that we are heading towards an unmanageable disaster by surpassing vital ecosystem thresholds (Butchart et al., 2010; MEA, 2005; Sanderson et al., 2002), and that we are running out of time to change our ways. Even the most energetic global initiatives taken today might not be sufficient to prevent the notable impoverishment and eventual collapse of life on Earth.

13. Case study: Management assessment of the protected areas of the Autonomous Region of Madrid, Spain

From 2008 to 2011, we developed the System for the Integrated Assessment of Protected Areas (SIAPA) in the Institute of Economics, Geography and Demography of the Spanish National Research Council (Rodríguez-Rodríguez & Martínez-Vega, in press). We applied the SIAPA to the ten PAs of a paradigmatically unsustainable region: the Autonomous Region of Madrid, Spain (Jiménez et al., 2005). In this epigraph, we show the part of the study's results of the implementation of the SIAPA (Rodríguez-Rodríguez & Martínez-Vega, under review) dealing with management of the ten PAs assessed.

The Autonomous Region of Madrid is a Spanish region of slightly over 8,000 km^2 in the centre of Spain (Figure 6). It has the highest per-capita income in Spain. It also has a rich natural and cultural patrimony that is jeopardized by the implementation of an intensive resource-consuming development model (Mata et al., 2009; Naredo & Frías, 2005), which has led to an increase in economic standards, but also to massive residential and infrastructure developments throughout the region in the past 20 years (Fernández-Muñoz, 2008; Gago et al., 2004).

The protected areas (PAs) of the region of Madrid face numerous pressures, mainly from massive visitor use and intensive land-use transformation (Rodríguez-Rodríguez, 2008) in addition to more general threats arising from global change, such as climate change (Araújo et al., 2011). These unsustainable development dynamics raise concern that effective long-term conservation of the goods and services provided by the region's ecosystems cannot be achieved despite the fact that up to 46 % of its territory is under some kind of protection regime (Mata et al., 2009).

The ten PAs selected for this study are shown in Table 2 and Figure 6.

Protected area	Abbreviation	Area (ha)	Designation year	MAI
Peñalara Natural Park	Peñalara NP	11,637	1990	1.6
Cuenca Alta del Manzanares Regional Park	Cuenca Alta RP	52,796	1985	1.2
Sureste Regional Park	Sureste RP	31,550	1994	1.1
Curso Medium del Río Guadarrama y su entorno Regional Park	Guadarrama RP	22,116	1999	1.1
Pinar de Abantos y Zona de la Herrería Picturesque Landscape	Pinar Abantos y Herrería PL	1,538	1961	0.6
Natural Site of National Interest of Hayedo de Montejo de la Sierra	NSNI Hayedo Montejo	250	1974	1.1
El Regajal-Mar de Ontígola Natural Reserve	Regajal-Ontígola NR	629	1994	0.6
Laguna de San Juan Fauna Refuge	Laguna San Juan FR	47	1991	0.7
Natural Monument of National Interest of Peña del Arcipreste de Hita	NMNI Peña Arcipreste	3	1930	0.0
Preventive Protection Regime of Soto del Henares	PPR Soto Henares	332	2000	0.3

MAI: Value of the Management Index of the SIAPA (in a 0 – 2 point standard scale)

Table 2. Main characteristics of the ten protected areas assessed

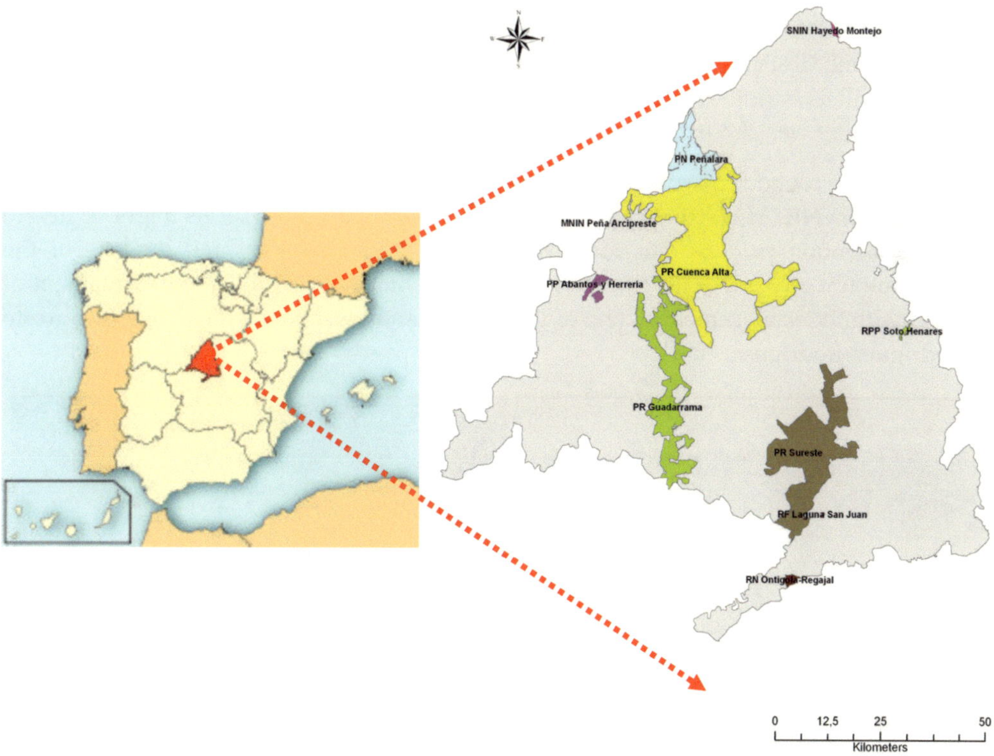

Figure 6. Location of the Autonomous Region of Madrid and the ten protected areas included in the study

The 12 management indicators against each PA was assessed are shown in Table 3

Category	Indicator
Management	Degree of characterization of the protected area
	Degree of fulfillment of management objectives
	Evolution of feature(s) for which the protected area was designated
	Existence of sufficient management staff
	Evolution of investment in the protected area
	Effectiveness of public participation bodies
	Production and distribution of an annual report on activities and outcomes
	Easiness to identify the protected area
	Public use infrastructure
	Existence of environmental education and volunteering activities
	Sanctioning procedures
	Monitoring activities

Table 3. Management indicators included in the SIAPA

Despite the importance of management for the effective conservation of PAs (Hockings et al., 2006; Pomeroy et al., 2005), the Management Index (MAI) did not reach a minimum desirable value for the PAs of the Autonomous Region of Madrid: 0.7 / 2 points. Only Peñalara NP had "Adequate" management. Whereas four PAs, including the three regional parks and the NSNI Hayedo Montejo, had a "Moderate" management valuation, the other five PAs had all "Deficient" valuations. The lowest value was for NMNI Peña Arcipreste (MAI = 0), as all its indicators scored 0 or lacked any information to be valuated. The MAI of PPR Soto Henares was also extremely low (MAI = 0.3).

Low scores were found for two PAs which had a director at the moment of the assessment: Regajal-Ontígola NR (MAI = 0.6) and Laguna San Juan FR (MAI = 0.7). This might be due to the scarce attention paid by the administration to these two PAs, and because of the excessive amount of work of the director, who had to make compatible the management of both PAs with the management of one of the biggest and most conflicting PAs: the Sureste RP (Rodríguez-Rodríguez, 2008).

Figure 7. Fauna observation installation in Sureste Regional Park (left figure) and information panels at the entrance of El Regajal-Mar de Ontígola Nature Reserve (right figure). Poor management and inadequate visitor behaviour result in the deterioration of PAs and their infrastructure

At least 2 of the 10 PAs assessed can be deemed "paper parks", as no management activities are performed there: NMNI Peña Arcipreste and PPR Soto Henares. Other 2 PAs (FR Laguna San Juan and El Regajal-Mar de Ontígola NR) enjoy only sporadic management activities, including surveillance. All these PAs have, at best, passive management responding reactively to management and conservation needs. Another PA (PP Pinar Abantos) has regular surveillance as the only habitual management activity. The other 5 PAs (4 parks and the NSNI Hayedo Montejo) do enjoy diverse active management regularly.

The indicators that scored highest for the ten PAs were the "existence of sufficient management staff", "evolution of investment in the PA", and "existence of environmental education and volunteering activities", all with a mild 1.1 score over a maximum of 2 points.

In contrast, the indicators that scored lowest were the "effectiveness of public participation bodies" (0.0 points) and the "easiness to identify the PA" (0.4 points). Additionally, two of the 12 indicators assessed under "Management" could not be assessed due to lack of rough data: "evolution of the feature(s) for which the PA was designated", and "sanctioning procedures".

A set of ultimate reasons could explain the poor quality of management of the PAs of the Autonomous Region of Madrid, as mentioned by Nolte et al. (2010) and Pomeroy et al. (2005): competence and information dispersal among different administrative units; poor coordination among these units on policies and activities related to PAs; shortage of human and material resources devoted to management; deficient scientific knowledge; and weak institutional support to biodiversity conservation policies.

In contrast to what we had expected, the MAI was not correlated with any other index used in the study, although it might have some degree of relatedness with the Planning Index (Rodríguez-Rodríguez & Martínez-Vega, in press). The non-significant, low correlation between the MAI and the State of Conservation Index and the non-significant, moderate correlation between the MAI and the Sustainability (Effectiveness)[5] Index suggest that management is not as determinant a factor for the effective conservation of PAs.

Although management might not be as fundamental a factor for the effective long-term conservation of PAs as it had been stated, it remains nevertheless one of the most important components of PA effectiveness. The importance of management increases in contexts of numerous and intensive inner and outer pressures on PAs, as it happens in the Autonomous Region of Madrid. Thus, much more effort will be needed regarding sustainable territorial management at different scales in this urban, densely-populated region to effectively safeguard its valuable natural and cultural resources for future generations (Mata et al., 2009; Rodríguez-Rodríguez, 2012b).

Author details

David Rodríguez-Rodríguez
World Commission on Protected Areas (IUCN-WCPA), Spain

---

[5] See Chapter 7

14. References

Alcaraz-Segura, D., Cabello, J., Paruelo, J.M., & Delibes, M. (2009). Use of Descriptors of Ecosystem Functioning for Monitoring a National Park Network: A Remote Sensing Approach. *Environmental Management* 43, 1, 38-48.

Araújo, M.B., Alagador, D., Cabeza, M., Nogués-Bravo, D. & Thuiller, W. (2011). Climate change threatens European conservation areas. *Ecology Letters* 14, 5, 484-492.

Barrado, D. A. (1999). *Actividades de ocio y recreativas en el medio natural de la Comunidad de Madrid.*

La ciudad a la búsqueda de la naturaleza [Nature leisure and recreacional activities in the Autonomous Region of Madrid. The city in search of nature]. Consejería de Medio Ambiente, Comunidad de Madrid, Madrid, Spain.

Beier, P. & Noss, R.F. (1998). Do habitat corridors provide connectivity? *Conservation Biology* 12, 6, 1241-1252.

Bertzky, M. & Stoll-Kleemann, S. (2009). Multi-level discrepancies with sharing data on protected areas: What we have and what we need for the global village. *Journal of Environmenal Management* 90, 1, 8-24.

Bonham, C.A., Sacayon, E., & Tzi, E. (2008). Protecting imperilled "paper parks"; potential lessons from the Sierra Chinajá, Guatemala. *Biodiversity Conservation* 17, 1581-1593.

Boitani, L., Falcucci, A., Maiorano, L. & Rondinini, C. (2007). Ecological Networks as Conceptual Frameworks or Operational Tools in Conservation. *Conservation Biology* 21, 6, 1414-1422.

Borrini-Feyerabend, G., Kothary, A., & Oviedo, G. (2004). *Indigenous and Local Communnities and Protected Areas: Towards Equity and Enhanced Conservation.* IUCN. Gland, Switzerland and Cambridge, UK.

Brunckhorst, D. (2000). Bioregional Planning. Resource management beyond the new millennium. Routledge, London.

Burel, F. & Baudry, J. 2003. Landscape Ecology. Concepts, Methods and Applications. Science Publishers INC. Enfield, USA.

Butchart, S.H.M., Walpole, M., Collen, B., van Strien, A., Scharlemann, J.P.W., Almond, R.E.E., Baillie, J.E.M., Bomhard, B., Brown, C., Bruno, J., Carpenter, K.E., Carr, G.M., Chanson, J., Chenery, A.M., Csirke, J., Davidson, N.C., Dentener, F., Foster, M., Galli, A., Galloway, J.N., Genovesi, P., Gregory, R.D., Hockings, M., Kapos, V., Lamarque, J.-F., Leverington, F., Loh, J., McGeoch, M.A., McRae, L., Minasyan, A., Morcillo, M.H., Oldfield, T.E.E., Pauly, D., Quader, S., Revenga, C., Sauer, J.R., Skolnik, B., Spear, D., Stanwell-Smith, D., Stuart, S.N., Symes, A., Tierney, M., Tyrrell, T.D., Vié, J.C. & Watson, R. (2010). Global biodiversity: indicators of recent declines. *Science* 328, 5982, 1164-1168.

CBD. Convention on Biological Diversity. (1992). Available from: http://www.cbd.int/convention/about.shtml

Chape, S., Spalding, M., & Jenkins, M.D. (2008). *The World's Protected Areas: Status, Values and Prospects in the 21st Century.* UNEP World Conservation Monitoring Centre, University of California Press, Berkeley, USA.

Davis, J.B. (Ed.) (2001). Paper Parks: why they happen and what can be done to change them? MPA News. International *News and Analyses on Marine Protected Areas* 2, 11, 1-6. Available from: http://depts.washington.edu/mpanews/MPA20.pdf

Dudley, N. (Ed.) (2008). *Guidelines for Applying Protected Area Management Categories.* IUCN, Gland, Switzerland.

Dudley, N., Stolton, S., Belokurov, A., Krueger, L., Lopoukhine, N., MacKinnon, K., Sandwith, T., & Sekhran, N. (2010). *Natural Solutions. Protected areas helping people cope with climate change.* IUCN-WCPA, TNC, UNDP, WCS, The World Bank and WWF. Gland, Switzerland, Washington DC and New York, USA.

Emerton, L., Bishop, J., & Thomas, L. (2006). *Sustainable Financing of Protected Areas: A global review of challenges and options.* IUCN. Gland, Switzerland, and Cambridge, UK.

Ervin, J. (2003a). *Rapid Assessment and Prioritization of Protected Area Management (RAPPAM) Methodology.* WWF, Gland, Switzerland.

Ervin, J. (2003b). Protected area assessment in perspective. *BioScience* 53, 9, 819-822.

Farrell, T. A. & Marion, J. L. (2001). Identifying and assessing ecotourism visitor impacts at eight protected areas in Costa Rica and Belize. *Environmental Conservation* 28, 3, 215-225.

Farrell, T.A. (2002). The Protected Area Visitor Impact Management (PAVIM) Framework: a simplified process for making management decisions. *Journal of Sustainable Tourism* 10, 31-51.

Fernández-Muñoz, S. (2008). Participación pública, gobierno del territorio y paisaje en la Comunidad de Madrid [Public participation, territorial governance and landscape in the Autonomous Region of Madrid]. *Boletín de la A.G.E* 46, 97-119.

Fraser, E.D.G., Dougill, A.J., Mabee, W.E., Reed, M., & McAlpine, P. (2006). Bottom up and top down: Analysis of participatory processes for sustainability indicator identification as a pathway to community empowerment and sustainable environmental management. *Journal of Environmental Management* 78, 2, 114-127.

Gago, C., Serrano, M., & Antón, F.J. (2004). Repercusiones de las carreteras orbitales de la Comunidad de Madrid en los cambios de usos del suelo [Land-use impacts of orbital roads of the Autonomous Region of Madrid]. *Anales de Geografía* 24, 145-167.

Gaston, K.J., Charman, K., Jackson, S.F., Armsworth, P.R., Bonn, A., Briers, R.A., Callaghan, C.S.Q., Catchpole, R., Hopkins, J., Kunin, W.E., Latham, J., Opdam, P., Stoneman, R., Stroud, D.A. & Tratt, R. (2006). *The* ecological effectiveness of protected areas: the United Kingdom. *Biological Conservation* 132, 1, 76-87.

Hillery, M., Nancarrow, B., Griffin, G., & Syme, G. (2001). Tourist perception of environmental impact. *Annals of Tourism Research* 28, 853-867.

Hockings, M., Stolton, S, & Dudley, N. (2000). *Evaluating effectiveness: a framework for assessing the management of protected areas.* IUCN. Gland, Switzerland, and Cambridge, UK.

Hockings, M., Stolton, S., Leverington, F., Dudley, N., & Courrau, J. (2006). *Evaluating effectiveness. A framework for assessing management effectiveness of protected areas.* (Second Ed.) IUCN, Gland, Switzerland, and Cambridge, UK.

Ioja, C.I., Patroescu, M., Rozylowicz, L., Popescu, V.D., Verghelet, M., Zotta, M.I., & Felciuc, M. (2010). The efficacy of Romania's protected areas network in conserving biodiversity. *Biological Conservation* 143, 11, 2468-2476.

Jameson, S.C., Tupper, M.H., & Ridley, J.M. (2002). The three screen doors: *Can* marine 'protected' areas be effective? *Marine Pollution Bulletin* 44, 1177-1183.

Jenkins, C.N. & Joppa, L. (2009). Expansion of the global terrestrial protected area system. *Biological Conservation* 142, 10, 2166-2174.

Jiménez, L., Prieto, F., Riechmann, J., & Gómez-Sal, A. (2005). *Sostenibilidad en España 2005. Informe de Primavera [Sustainability in Spain 2005. Spring Report]*. Observatorio de la Sostenibilidad de España, Alcalá de Henares, Spain.

Jongman, R. (2002). Homogenisation and fragmentation of the European landscape: ecological consequences and solutions. *Landscape and Urban Planning* 58, 211-221.

Kerry, R., Paavola, J., Cooper, P., Farber, S., Jessamy, V., & Georgiu, S. (2003). Valuing nature: lessons learned and future research directions. *Ecological Economics* 46, 3, 493-510.

Koplylova, S.L. & Danilina, N.R. (Eds.). (2011). *Protected Area Staff Training: Guidelines for Planning and Management*. IUCN, Gland, Switzerland.

Leverington, F., Lemos, K., Courrau, J., Pavese, H., Nolte, C., Marr, M., Coad, L., Burguess, N., Bomhard, B. & Hockings, M. (2010). *Management effectiveness evaluation in protected areas – a global study*. (Second Edition). University of Queensland, Brisbane, Australia.

Mallarach, J.M., Germain, J., Sabaté, X., & Basora, X. (2008). *Protegits de fet o de dret? Primera avaluació del sistema d'espais naturals protegits de Catalunya [Actually or legally protected? First assessment of the system of protected areas of Catalonia]*. Institució Catalana d'Història Natural. Available from: http://ichn.iec.cat/Avaluaci%C3%B3%20d%27espais.htm

Mata, R., Galiana, L., Allende, F., Fernández, S., Lacasta, P., López, N., Molina, P. & Sanz, C. (2009). Evaluación del paisaje de la Comunidad de Madrid: de la protección a la gestión territorial [Evaluation of the landscape of the Autonomous Region of Madrid: from protection to territorial Management]. *Urban* 14, 34-57.

McDonald, R.I. & Boucher, T.M. (2011). Global development and the future of the protected area strategy. *Biological Conservation* 144, 1, 383-392.

McKercher, B. & Weber, K. (2008). Rationalising inappropriate behaviour at contested sites. *Journal of Sustainable Tourism* 16, 4, 369-385.

MEA. Millennium Ecosystem Assessment. (2005). *Ecosystems and Human Well-Being: Synthesis*. Island Press. Washington DC.

Metera, E., Sakowsky, T., Sloniewsky, K. & Romanowicz, B. (2010). Grazing as a tool to maintain biodiversity in grasslands. A review. *Animal Science Papers and Reports* 28, 4, 315-334.

Mora, C. & Sale, P.F. (2011). Ongoing global biodiversity loss and the need to move beyond protected areas: a review of the technical and practical shortcomings of protected areas on land and sea. *Marine Ecology Progress Series* 434, 251-266.

Morales, J. (2001). *Guía Práctica para la Interpretación del Patrimonio. El arte de acercar el legado natural y cultural al público visitante [Practice Guide for the Interpretation of Patrimomy. The art of approaching natural and cultural legacies to visitors]*. Junta de Andalucía, Seville, Spain.

Múgica, M. & Gómez-Limón, J. (Coords.). (2002). *Plan de Acción para los espacios naturales protegidos del Estado español [Action Plan for the protected areas of the Spanish State]*. Fundación Francisco González Bernáldez, Madrid, Spain.

Múgica, M., De Lucio, J. V., Martínez-Alandi, C., Sastre, P., Atauri-Mezquida, J. A. & Montes, C. (2002). *Integración territorial de espacios naturales protegidos y conectividad ecológica en paisajes mediterráneos [Territorial integration of protected areas and ecological connectivity in Mediterranean landscapes]*. Consejería de Medio Ambiente, Junta de Andalucía, Seville, Spain.

Naredo, J.M. & Frías, J. (2005). *Desarrollo: la síntesis del "desarrollo sostenible" con especial referencia a la Comunidad de Madrid [Development: the síntesis of "sustainable development" with special referente to the Autonomous Region of Madrid].* In Sánchez-Herrera, F., pp. 7-38. *Cuartas Jornadas Científicas del Parque Natural de Peñalara y del Valle de El Paular. Conservación y desarrollo socioeconómico en Espacios Naturales Protegidos [Fourth Scientific Meeting of Peñalara Natural Park and El Paular Valley. Conservation and Socioeconomic Development in Protected Areas].* Consejería de Medio Ambiente y Ordenación del Territorio, Comunidad de Madrid, Madrid, Spain.

Naughton-Treves, L., Buck, M., & Brandon, K. (2005). The Role of Protected Areas in Conserving Biodiversity and Sustaining Local Livelihoods. *Annual Review of Environmental Resources* 30, 219-252.

Nolte, C., Leverington, F., Kettner, A., Marr, M., Nielsen, G., Bomhard, B., Stolton, S., Stoll-Kleemann, S., & Hockings, M. (2010). *Protected Area Management Effectiveness Assessments in Europe. A review of application, methods and results*. University of Greifswald, Germany.

Olson, D. M., Dinerstein, E., Wikramayanake, E. D., Burgess, N. D., Powell, G. V. N., Underwood, E. C., D' Amico, J. A., Itoua, I., Strand, H. E., Morrison, J. C., Loucks, C. J., Allnutt, T. F., Ricketts, T. H., Kura, Y., Lamoreux, J. F., Wettengel, W. W., Hedao, P. & Kassem, K. R. (2001). Terrestrial Ecoregions of the World: A New Map of Life on Earth. *BioScience* 51, 11, 933-938.

Parrish, J.D., Braun, D.P., & Unnasch, R.S. (2003). Are we conserving what we say we are? Measuring the ecological integrity within protected areas. *BioScience* 53, 9, 851-860.

Phillips, A. (Ed.). (2000). *Financing Protected Areas. Guidelines for Protected Area Managers*. IUCN, Gland, Switzerland and Cambridge, UK.

Pomeroy, R.S., Parks, J.E. & Watson, L.M. (2005). *How is your MPA doing? A methodology for evaluating the management effectiveness of marine protected areas*. IUCN, Gland, Switzerland, and Cambridge, United Kingdom.

Pressey, R. L., Cabeza, M., Watts, M. E., Cowling, R. M. & Wilson, K. A. (2007). Conservation planning in a changing world. *Trends in Ecology and Evolution* 22, 11, 583-592.

Pullin, A. (2002). *Conservation Biology*. Cambridge University Press, Cambridge, UK.

Radeloff, V.C., Stewart, S.I., Hawbaker, T.J., Gimmi, U., Pidgeon, A.M., Flather, C.H., Hammer, R.B. & Helmers, D.P. (2010). Housing growth in and near United States protected areas limits their conservation value. *PNAS* 107, 2, 940-945.

Rodríguez-Rodríguez, D. (2008). *Los espacios naturales protegidos de la Comunidad de Madrid. Principales amenazas para su conservación [The protected areas of the Autonomous Region of Madrid. Main threats to their conservation].* Editorial Complutense, Madrid, E-Book, Available from: http://www.ucm.es/BUCM/ecsa/36254.php?id=187

Rodríguez-Rodríguez, D. (2009). Mitigación de los impactos del turismo en espacios naturales protegidos y mejora de su financiación a través de medidas económicas. El caso de la Comunidad de Madrid [Tourism impact mitigation and enhanced financing of protected areas through economic measures. The case of the Autonomous Region of Madrid]. *Boletín de la A.G.E.* 50, 217-238.

Rodríguez-Rodríguez, D. (2012a). Perception use and valuation of protected areas by local populations in an economic crisis context. *Environmental Conservation* 39, 2, 162-171.

Rodríguez-Rodríguez, D. (2012b). Integrated Networks. A territorial planning proposal for biodiversity conservation in urban, densely populated regions. The case of the Autonomous Region of Madrid, Spain. *Journal of Environmental Planning and Management* 55, 5, 667-683.

Rodríguez-Rodríguez, D. & Martínez-Vega, J. (In press). Proposal of a system for the integrated and comparative assessment of protected areas. *Ecological Indicators,* doi: 10.1016/j.ecolind.2012.05.009

Rodríguez-Rodríguez, D & Martínez-Vega, J. (Under review). Results of implementation of the System for the Integrated Assessment of Protected Areas (SIAPA) to the protected areas of the Autonomous Region of Madrid (Spain). *Ecological Indicators.*

Sanderson, E.W., Jaiteh, M., Levy, M.A., Redford, K.H., Wannebo, A.V. & Woolmer, G. (2002). The Human Footprint and the Last of the Wild. *BioScience* 52, 10, 891-904.

Sandwith, T., Shine, C., Hamilton, L. & Sheppard, D. (2001). *Transboundary Protected Areas for Peace and Co-operation.* IUCN, Gland, Switzerland and Cambridge, UK.

Spangenberg, J.H. (2002). Environmental space and the prism of sustainability: frameworks for indicators measuring sustainable development. *Ecological Indicators* 2, 3, 295-309

Thomas, L. & Middleton, J. (2003). *Guidelines for Management Planning of Protected Areas.* IUCN, Gland, Switzerland, and Cambridge, UK.

Tisdell, C. (2001). *Tourism Economics, the Environment and Development. Analysis and Policy.* Edward Elgar Publishing, Cheltenham, UK.

TNC. The Nature Conservancy. (2001). *Conservation by Design: A Framework for Mission Success.* Arlington, Virginia, USA. Available from: http://www.fws.gov/southeast/grants/pdf/cbd_en.pdf

Tzoulas, K., Korpela, K., Venn, S., Yli-Pelkonen, V., Kazmierczak, A., Niemela, J. & James, P. (2007). Promoting ecosystem and human Health in urban areas using Green Infrastructure: A literature review. *Landscape and Urban Planning* 81, 3, 167-178.

UNEP-WCMC. United Nations Environment Programme-World Conservation Monitoring Centre. (2008). *State of the world`s protected areas 2007. An annual review of global conservation progress.* UNEP-WCMC, Cambridge, UK.

UNESCO. (2002). *Biosphere reserves: Special places for people and nature.* UNESCO, Paris, France.

VVAA. (2000). *La conservación de la naturaleza: aspectos clave y retos de futuro. Seminarios de Conservación de la Naturaleza (1996-1999) [Nature conservation: key aspects and future challenges.* Nature Conservation Seminars (1996-1999)] Serie Documentos nº 30, Centro de Investigación "Fernando González Bernáldez", Soto del Real, Madrid, Spain.

Walpole, M.J., Goodwin, H. J. & Ward, K.G.R. (2001). Pricing policy for tourism in protected areas: lessons from Komodo National Park, Indonesia. *Conservation Biology* 15, 1, 218–227.

WCED. World Commission on Environment and Development. (1987). *Our Common Future.* Oxford University Press, Oxford, UK.

Witte, F., Goldschsmidt, T., Goudswaard, P.C., Ligtvoet, W., Van Oijen, M.J.P. & Wanink, J.H. (1991). Species Extinction and Concomitant Ecological Changes in Lake Victoria. *Netherlands Journal of Zoology* 42, 2, 214-232.

Zamora, R. (2010). Las Áreas Protegidas como Observatorios del Cambio Global [Protected Areas as Observatories of Global Change]. *Ecosistemas* 19, 2, 1-4.

Chapter 3

Managing the Wildlife Protected Areas in the Face of Global Economic Recession, HIV/AIDS Pandemic, Political Instability and Climate Change: Experience of Tanzania

Jafari R. Kideghesho and Tuli S. Msuya

Additional information is available at the end of the chapter

http://dx.doi.org/10.5772/51335

1. Introduction

World Conservation Union [1] defines a protected area as: "An area of land and/or sea especially dedicated to the protection and maintenance of biological diversity, and of natural and associated cultural resources, and managed through legal or other effective means." Protected areas are intended to meet one or more of the following purposes: scientific research; education; wilderness protection; preservation of species and genetic diversity; maintenance of environmental services; protection of specific natural and cultural features; tourism and recreation; sustainable use of resources from natural ecosystems; and maintenance of cultural and traditional attributes[1]. Each of these management purposes is related to a category of protected areas i.e., groups of protected areas assigned to cater for specific purpose or objective.

Along with other benefits associated with protection and maintenance of biodiversity, justification for the establishment of protected areas in many developing countries indicates a bias on economic rather than ecological benefits. Many protected areas are established because of their economic potential. They generate significant multiplier effects across a national economy, and offer considerable economic value to the livelihoods of the poorest and most vulnerable sectors of society. They create investment opportunities and employment. Essentially, protected areas are recognized as important vehicle towards poverty reduction and sustainable development [2,3]. The most important avenue through which protected areas contribute significantly to local and national economy is through tourism industry. Protected areas are cherished as the key tourist destinations offering a variety of attractions to domestic and international visitors. They are also important hunting grounds catering for international tourists and residents. Essentially, both consumptive and non-consumptive forms of tourism are recognized as important economic engine and a development strategy for many developing countries [4-6]. It is the largest in terms of contribution to the global GDP and second, after agriculture, in provision of employment [7].

The ability of protected areas to provide multiple benefits to humanity is, however, compromised by numerous factors causing overexploitation of species, habitat destruction, pollution and introduction of exotic species. Globally, there is a growing trend of biodiversity loss and an increase of species threatened with extinction. For example, of the 44,838 species included in the 2008 IUCN Red List database, about 17,000 (38%) were threatened with extinction. Comparison of the IUCN Red Lists for 1996 and 2008 indicates that the number of species threatened with extinction had grown [8,9]. In Southern Africa, poachers and organized criminal gangs, who supply the lucrative international ivory and rhino-horn markets, are reported to have caused significant negative ecological impacts on rhino and elephant. According to report, many parks in South Africa were experiencing a growing trend of rhino poaching. For example, between 2001 and 2006 about 70 rhinos were killed in Kruger National Park alone [10]. The most known and documented factors leading to these trends include human population growth, poverty, failure of conservation – as an alternative form of land use - to compete effectively with forms of land uses that are ecologically destructive, and inability of legal economic benefits from protected areas to offset the conservation related costs incurred by local communities through property damage, wildlife-related accidents and numerous opportunity costs.

Multiple benefits derived from the protected areas and growing threats facing them have prompted a dramatic increase of land under protection globally (Figure 1). Essentially, the protected areas are increasingly being acknowledged as the most effective tools for conservation of biodiversity – genes, species and ecosystems. The 2010 World Database on Protected Areas Annual Release [11] indicates that over 160,000 protected areas covering over 21 million square kilometres of land and sea have been established to date. Of these, terrestrial protected areas exceed 12% of the Earth's land area and marine protected areas occupy about 6% of the Earth's territorial seas. In recent years, the protected area coverage has been adopted as an indicator to measure the policy response to biodiversity loss in different countries. Efforts by governments and civil societies to conserve biodiversity are measured by the increased land and sea areas put under protection. The use of protected area coverage as an indicator is in line with the CBD's 2010 target of achieving a significant reduction of the rate of biodiversity loss [12].

The effectiveness of protected areas as the leading strategy in global efforts of stemming loss of biodiversity is, however, being challenged. It is argued that the effectiveness of the existing and the current pace of the establishment of the new protected areas can hardly reverse the current trends of biodiversity loss [14]. The deficiencies of the protected areas undermining their conservation goals include:

- The slow rate of expanding the protected areas to cope with the current threats of biodiversity;

- Inability of the protected areas to overcome all threats: The roles of protected areas can effectively mitigate the problems of species overexploitation and habitat loss, but has limited capacity to overcome other stressors, such as climate change, pollution, and invasive species;
- The increasing need for human development at the expense of wildlife habitats and species thus, creating conflicts with conservation goals;
- Insufficient size and connectivity of protected areas and, consequently, failure to sustain viable populations and allow exchange of genetic materials between individuals;
- Inadequate funding of the protected areas which undermines their effective management. Annual estimate for effective management of protected areas is $24 billion — four times the current expenditure of $6 billion [14].

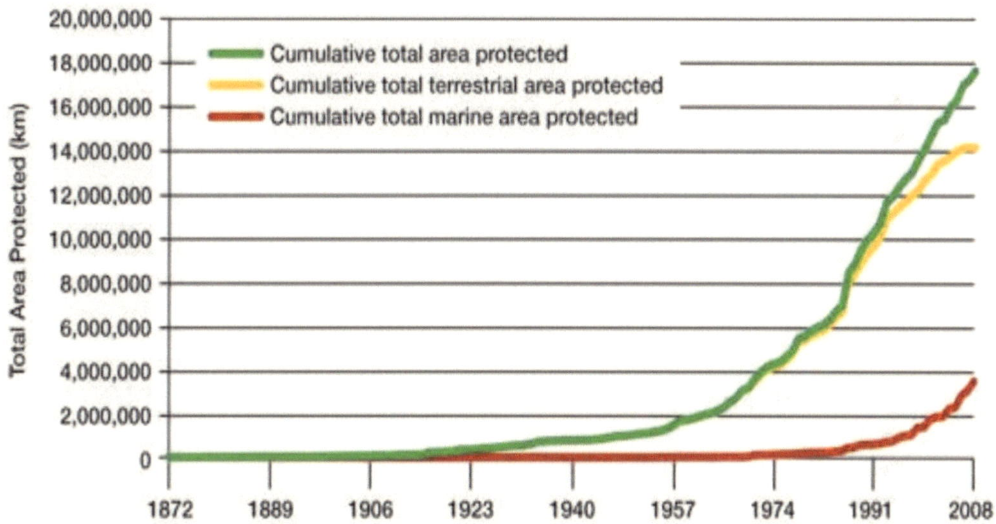

Figure 1. Growth in nationally designated protected areas from 1872 to 2008 (Graph excludes protected areas with unknown year of establishment). Source: [13].

The failure of protected areas in their conservation role is worsened by issues, which unfortunately are inadequately documented in literature as they have only emerged recently or they were existing but were not recognized as potential threats. Because of their freshness, their attention in conservation literature and policies has been minimal. This chapter seeks to examine these emerging issues in order to increase public awareness on impacts associated with these issues and stimulate feasible and sustainable interventions from different actors.

2. Framing the issue

Human population growth and poverty are regarded as the underlying causes for biodiversity loss in protected areas through overexploitation of natural resources, habitat destruction, introduction of exotic species and pollution. However, behind these causes, there are numerous factors determining their magnitudes and impacts on natural resources.

While the impacts of numerous factors on conservation and protected areas are well established in literature, the impacts for some have remained insufficiently documented, most likely because they have only recently emerged and/or recognized as threats to conservation. The factors whose impacts on conservation and protected areas are minimally acknowledged in literature include global economic recession, climate change, HIV/AIDS pandemic and civil wars. Global economic recession may generate poverty at, national and household levels and, consequently, affect the conservation sector and protected areas management by reducing funding and increasing human pressure on species and habitats. Similarly, HIV/AIDS pandemic may cause overexploitation of species and destruction of habitats when the victims remain with limited options to meet their livelihood strategies and medicinal needs. Impacts of climate change can be manifested through food insecurity and poverty, effects on species and habitats and worsening human-wildlife conflicts. Political instability cause poverty as people can hardly work to earn their living in a warfare environment. On the other hand, wars cause an influx of refugees and, therefore, contribute to human population growth. The high human population creates more demand for natural resources at the expense of species and habitats. In light of the scenarios mentioned here, it is apparent that these factors have notable ecological impacts on conservation sector and protected areas, in particular. It is, therefore, imperative that they are critically analyzed and brought to the attention of policy makers, conservation planners and public at large. Planning for protected areas should consider these factors as issues of urgency calling for special priority.

3. Wildlife conservation in Tanzania and establishment of wildlife protected areas

Tanzania's conservation history dates back to early 1890s when the German Administration enacted the first Wildlife Law in order to regulate hunting. The British Administration, which took over in 1920 following defeat of the Germans in the World War I, continued to make wildlife conservation a matter of priority. The British regime enacted the first comprehensive wildlife conservation legislation, the Game Preservation Ordinance of 1921. Pursuant to the provisions of this Ordinance, Serengeti was declared a partial Game Reserve in 1921 and elevated to a full one in 1929. The Selous Game Reserve was gazetted as the first game reserve in 1922. In 1921, the Game Department was established to administer the game reserves, enforce the hunting regulations and control the problem animals [15].

In gazetting the protected areas, precautions were taken by colonial administrators in Tanzania not to infringe on African rights as this could lead to political instability of the colony. However, pressure for more restrictive and prohibitive conservation laws along with setting aside more lands exclusively for conservation came from Europe, spearheaded by the London-based Society for Preservation of Flora and Fauna of the Empire (SPFFE) and other powerful conservation lobby. In 1930, the Society sent Major Richard Hingston to investigate the needs and potential for developing a nature protection programme in Southern and Central Africa. One of the recommendations by Hingston was based on formalizing a more restrictive category of protected areas (i.e. national parks). Serengeti,

Kilimanjaro and Selous were proposed as ideal for the purpose of creating national parks in Tanganyika [16, 17]. The main criterion employed to rate an area's suitability as a national park was assurance that the area was unsuitable for Europeans' economic activities such as mining, livestock keeping and crop production.

Hingston's recommendations provided a basis for agenda of the 1933 London Convention on wildlife. All signatories (including Tanganyika) were required to investigate the potentials of creating a system of national parks. Colonial administrators in Tanganyika remained adamant for seven years, a situation that caused serious accusations from Europe that the colony was the worst offender in encouraging slaughter of game by the natives. These pressures paved the way to the first Game Ordinance that gave the governor a mandate to declare any area a national park. The Ordinance, enacted in 1940 repealed the 1921 Ordinance. Serengeti National Park was established in 1940 but remained a 'park in the paper' until 1951 as there was weak enforcement of regulations and laws governing the national parks.

Restrictive and prohibitive laws made the four decades of conservation under British rule be manifested by conflicts and resentment from the natives. For example, the Maasai tribe in eastern Serengeti resented the proposed park boundaries through violence and sabotage/vandalism. Their retaliatory response involved spearing of rhinos, setting fires with malicious intent and terrorising civil servants [17]. The Ikoma tribe of western Serengeti declared daringly that they would kill any wildlife ranger who would attempt to stop them from hunting and obtaining resources from Serengeti National Park.

As Tanzania was about to attain her political independence, there was a hope among the local communities and a fear among the European conservationists. The natives perceived independence as an end to stringent conservation laws that infringed upon their customary rights [16]. The conservationists were worried that political independence would decolonize nature by terminating the conservation efforts, mainly because Tanzanians had low capacity to carry out managerial activities in protected areas [16]. However, conservationists' fear was dissuaded when the post-colonial government endorsed continuation of colonial conservation policies uncritically. Economic rather than ecological reasons justified this policy choice. The wildlife-based tourism was perceived as a vital economic engine and insurance in case of failure of other economic sectors such as agriculture and minerals and, therefore, the government was not ready to forego this option. Julius Nyerere, the first Tanzanian President, was quoted saying:

> *"I personally am not interested in animals. I do not want to spend my holidays watching crocodiles. Nevertheless, I am entirely in favour of their survival. I believe that after diamonds and sisal, wild animals will provide Tanganyika with its greatest source of income. Thousands of Americans and Europeans have the strange urge to see these animals"* [18]

It is because of economic potential that land under legal protection has dramatically expanded in the past 50 years of Tanzanian independence. Today, while 55% of 236 countries have less than 10% of their land areas under legal protection [11], Tanzania has gazetted about 30% and 15% of its terrestrial land area as wildlife and forest protected areas,

TANZANIA
Protected Areas
(not an authority on boundaries)
0 100 200 km
UGANDA
KENYA
RWANDA
BURUNDI
D.R. CONGO
ZAMBIA
MALAWI
MOZAMBIQUE
INDIAN OCEAN
Lake Victoria
Lake Tanganyika
Lake Rukwa
Lake Nyasa
Malagarasi
Pangani
Wami
Ruaha
Rufiji
Bukoba
Mwanza
Arusha
Moshi
Kigoma
Tabora
Tanga
Pemba
Zanzibar
Dar es Salaam
Dodoma
Morogoro
Mafia
Iringa
Mbeya
Songea
Lindi
Mtwara
IBANDA
RUMANYIKA
KIMISI
RUBONDO ISLAND
BURIGI
IKORONGO
GRUMETI
SERENGETI
NGORONGORO
MASWA
ARUSHA
KILIMANJARO
TARANGIRE
MKOMAZI
UMBA
SWAGA SWAGA
KIGOSI
MOYOWOSI
GOMBE
UGALLA RIVER
MAHALE MTS
SAADANI
PANDE
RUNGWE
KISIGO
RUAHA
USANGU
UDZUNGWA
MIKUMI
SELOUS
KILOMBERO
MAHENGE
KITULO PLATEAU
LUAFI
LUKWATI
MANJESI
LUKWIKA LUMESURE
LIPARAMBA
© Cartography by Mike Shand

Protected area	*Size (Km²)*	*Year established*	*Notes*
Nature Reserves (IUCN Category I)			
Amani	83.8	1997	Formed from 6 FRs: Kwamkoro, Kwamsambia, Mnyuzi Scarp, Amani Zigi, Amani East & West
Kilombero	1,345.1	2007	Formed by merging Matundu, Iyondo and West Kilombero Scarp FRs
Nilo	62.5	2007	Upgraded from a FR
Chome	142.83	Proposed	Was designated as FR in 1951
Magamba	87	Proposed	Notified as FR in 1942; was scheduled to be upgraded in April 2010.
Mkingu	233.9	Proposed	To include Nguru South and Mkindo FRs
Udzungwa Scarp	327.63	Proposed	Notified as FR in 1929
Uluguru	241.2	2009	Links 3 former FRs: Uluguru North and South and Bunduki)
National Parks (IUCN category II)			
Arusha	137	1960	Known as Ngurdoto Crater NP until 1967, expanded in 1973)
Gombe Stream	52	1968	
Jozani Chwaka Bay	50	2004	The only national park in Zanzibar Island
Katavi	4,471	1974	
Kilimanjaro	755	1973	World Heritage Site since 1987)
Kitulo	412.9	2005	
Lake Manyara	664	1960	Enlarged 2009: original size 330 km²)
Mahale Mountains	1613	1985	
Mikumi	3230	1964	Extension in 1975
Mkomazi	3 270	2008	Game Reserve since 1951
Ruaha	22000	1964	Expanded in 2009: original size 10 300 km²
Rubondo Island	240	1977	Game Reserve since 1965
Saadani	1,100	2005	Game Reserve since 1969
Serengeti	14 763	1951	Game Reserve since 1928; Biosphere Reserve and World Heritage Site since 1981

Tarangire	2 850	1970	
Ngorongoro CA	8260	1959	Game Reserve since 1928; Biosphere Reserve and World Heritage Site since 1981
Some Game Reserves			
Biharamulo	1,300	1959	
Burigi	2,200	1972	
Ibanda	294	1974	
Ikorongo Grumeti	3 300	1994	
Kijereshi	65.7	2001	
Kimisi	1,026.23	2002	
Liparamba	570.99	2000	
Kizigo	4 000	1982	
Lukwati	3,146	1997	
Lukwika/Lumesule	444	1995	
Maswa	2 200	1962	
Mkungunero	700	1996	
Mpanga- Kipengele	1,574.25	2002	
Msanjesi	210	1995	
Muhesi	2,000	1994	
Muhesi	2 000	1994	
Pande Forest	12	1994	
Rukwa	4,000	1995	
Rumanyika	245	1970	
Rungwa	9 000	1951	
Saadani	4,000	1995	Annexed to Ruaha National Park in 2009
Selous	50,000	1922	Word Heritage Site since 1982
Swagaswaga	871	1996	
Ugalla	5 000	1965	
Uwanda	5 000	1971	
Ramsar Sites			
Malagarasi-Moyovosi	32,500	2000	
Lake Natron Basin	2,250	2001	
Kilombero Valley Floodplain	7,967	2002	
Rufiji -Mafia-Kilwa Marine	5970	2004	

Source: [21].

Table 1. The major wildlife protected areas of Tanzania

respectively. According to 2005 World Database on Protected Areas, over 11% of protected areas in Tanzania was under IUCN category I and II, 26% under category III - V and 63% under category VI and others [19]. More protected areas have been gazetted or upgraded to higher categories since 2005 and, therefore, these figures do not reflect the recent changes.

At the independence there were only three national parks (Serengeti, Lake Manyara and Arusha); nine game reserves and Ngorongoro Conservation Area. Today the number has grown to 16 national parks comprising an area of over 42,000 km² (4.4% of the country's land surface: see Table 1 and Figure 1). Over 30 game reserves have been gazetted along with adoption of three new categories of protected areas. These categories are Ramsar Sites, Wildlife Management Areas (WMAs) and Nature Reserves. The four Ramsar sites cover about 5.5% of Tanzania's wetlands. The WMAs have emerged as a key option following the recognition by the Wildlife Policy of 1998 (revised in 2007) [15, 20] that the future of wildlife in Tanzania rests on the ability of wildlife to generate economic benefits to the rural communities who live alongside wildlife, and its ability to compete effectively with other forms of land uses which are ecologically destructive. WMAs are, therefore, established as one of the strategies for implementing community wildlife management in Tanzania. WMAs were first legally formalized through the WMA Regulations of 2002 (revised 2005) and are now formalized in the Wildlife Conservation Act of 2009. Currently 14 WMAs have been designated and 20 others are in the process. The designated WMAs and their locations in brackets include Burunge (Babati), Uyumbu (Tabora), Makao (Shinyanga), IKONA/Ikoma-Nata (Serengeti), MBOMIPA/Pawaga Idodi (Iringa), Mbarang'andu (Namtumbo – Ruvuma), Magingo (Liwale - Lindi), Enduiment (West Kilimanjaro), Ipole (Sikonge, Tabora), Nalika (Tunduru–Ruvuma), MUNGATA/Ngarambe Tapika (Rufiji), Wamimbiki (Morogoro & Bagamoyo), JUKUMU (Morogoro), Kimbanda (Namtumbo) and Chingoli (Tunduru).

4. Emerging issues in the management of wildlife protected areas in Tanzania

As pointed out earlier, protected areas are intended to meet a variety of management purposes in order to support human livelihood and development through provision of ecosystem goods and services in a sustainable way. However, numerous ecological, socio-economic and political factors tend to undermine this desire. Of these factors, are the traditional ones, which are sufficiently covered in literature and, those which have emerged just recently. The latter are underrepresented in literature and, therefore, their inclusion in policies and management plans for many protected areas are lacking. Four of these emerging factors include global economic recession, HIV/AIDS pandemic, climate change and political instability. This section examines these issues by pointing out their potential impacts on the management of wildlife protected areas. Relevant examples are drawn from different protected areas of Tanzania.

4.1. Global economic recessions

A global economic recession is a period of general economic decline; typically defined as a decline in GDP for two or more consecutive quarters. A recession is typically accompanied by a drop in the stock market, an increase in unemployment, and a decline in the housing market. The World Bank's Global Development Finance report [22] placed Tanzania along with Ghana, Mali and Mozambique at relatively more risk to shocks associated with global economic recession compared to other African countries. This is due to considerable share of foreign owned banks and heavy reliance of economies on foreign direct investment in these countries [23]. Global economic recession may bear direct and indirect undesirable impacts in the management of wildlife protected areas by exacerbating poverty to people and, therefore, increasing pressure on natural resources. The recession also affects tourism sector which is the main source of revenues required to run the protected areas. Protected areas may also suffer through reduced support from donors, who fund different conservation programmes. Poor funding of the protected areas, consequently, undermine numerous activities and operations such as ecological monitoring, conservation education for local communities and law enforcement. These impacts are briefly discussed below.

4.1.1. Increased incidences of poverty and vulnerability

The financial recession in poor countries fuels incidences of poverty and vulnerability to individuals and at the national level. For example, reports indicate that Tanzania experienced significant loss economically due to 2007-2009 Global Financial Recession, despite the fact that it lasted for a very short time. The country's economy was projected to grow by 8% in 2009 but the crisis lowered this projection to 5% and 6% for 2009 and 2010. The

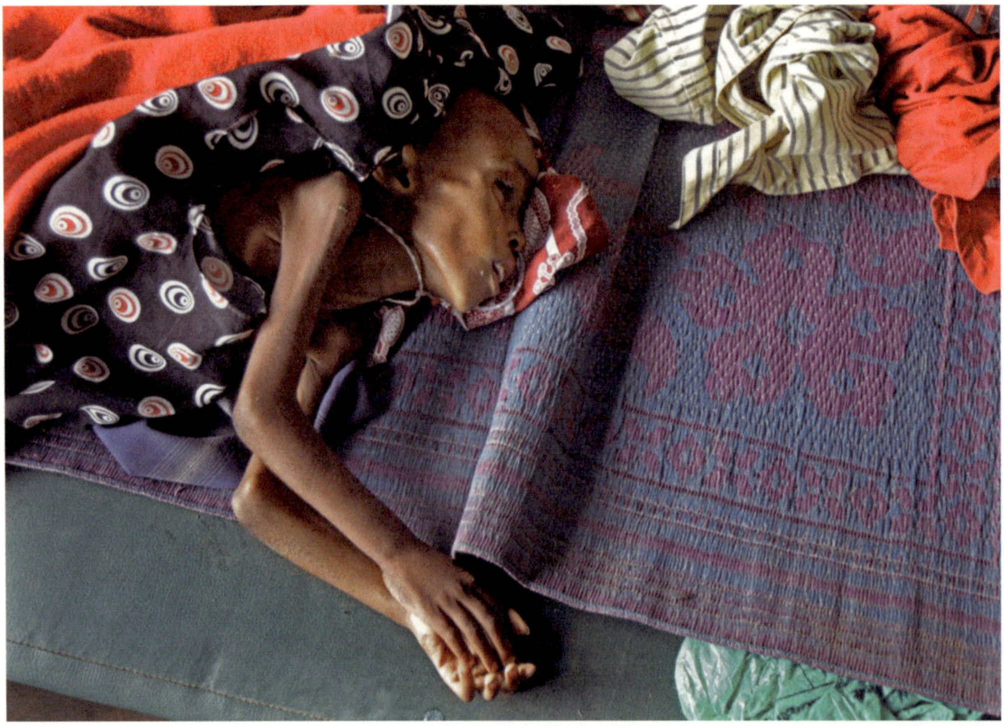

Figure 2. The need to feed the family prompts poaching of wildlife from protected areas

country estimated a loss of about US$255 million from domestic income occasioned by the recession [24]. The financial crisis also affects decisions to review the minimum wage for both public and private sector workers. The decline of national economy is translated to several development factors including natural resources and tourism industry (24-26).

Experience indicates that household poverty detrimentally affect the protected areas and natural resource base. Reduced ability of households to improve on existing livelihood strategies, forces them to adopt the coping strategies that are unsustainable and ecologically destructive. For example, because of poverty peasants can barely afford to purchase and use agricultural inputs to increase crop production in a piece of land. Food insecurity and income poverty resulting from this scenario may lead to conversion of more wildlife habitats into croplands as well as killing of wild animals for protein as evidenced in Serengeti National Park and adjacent protected areas [27, 28]. Household poverty also limits people from access and use of electricity as a source of energy, thus making wood fuel (firewood and charcoal) the most dominant and reliable source of energy for cooking and heating, both in urban and rural areas [27]. In order to meet an increased demand for wood fuel, wildlife habitats and other critical wildlife areas are subjected to deforestation.

4.1.2. Reduced tourism revenues

Wildlife protected areas in Tanzania and elsewhere in Africa rely on external sources for their survival. These sources are international tourism and donors. Nature-based tourism accounts for about 95% of internal revenues for Tanzania National Parks (TANAPA) and Ngorongoro Conservation Area Authority (NCAA) - the government organizations managing the national parks and Ngorongoro Conservation Area, respectively [29]. This implies that the wildlife protected areas can hardly survive without tourism. The revenues generated through tourism industry are required for conservation work. International tourism, contrary to domestic tourism, commands higher priority because the latter is less developed and, therefore, its contribution to economy is insignificant.

Global economic recessions have many and far-reaching impacts on the performance of the tourism sector and, therefore, management of the protected areas. This is epitomized by the global economic crisis that started in late 2007 in the United States of America and in some European countries. The growth of about 7.0% in international tourist arrivals from 719,030 in 2007 to 770,376 in 2008 was relatively small in comparison to that of between 2006 and 2007, which was about 12.0% [25]. This relative decrease in growth rate for international tourists to Tanzania between the two seasons reduced the earlier projected earnings and interfered with employment. The conservation agencies – TANAPA and NCAA- were compelled to cut down their annual expenditures in 2008/2009 [30, 31] which among other expending areas included resource protection.

Tourism industry recorded a decline in revenue by about 18% in 2008 and it was predicted that by 2009 decline would be 30% [26]. The crisis further impacted on the tourism value chain including travel agents, transporters (taxis, buses, car rentals, and Safari/tour operators), hotels, restaurants and camping sites as 60% cancellation of bookings was reported. The reduced number of tourists affected employment in hotels, restaurants and camping sites and their suppliers of food, beverages, laundry and utilities [26]. As a result of this crisis, some hotels and tour operators carried out their business below their capacities and others contemplated to close and lay off employees [25]. Laying off of the employees in different sectors may present another problem to protected areas as the redundant employees may resort to pursuing illegal activities such as poaching and unsustainable use of natural resources (e.g., fuel wood) in order to survive.

4.1.3. Reduced donor support

As is the case with many public sector budgets in Tanzania, external donor countries and development partners contribute notably to funding of the protected areas. Many conservation programmes and projects depend on funding from international agencies and other foreign donors. However, despite these efforts, there had been a global concern that funding of protected areas is inadequate [32]. It is, therefore, apparent that global economic recessions worsen the situation as many international donors can hardly honour their commitments to different projects/programmes including those related to conservation of biodiversity and, management of protected areas, in particular [25, 33, 34]. Essentially, it is unlikely for donors to pay adequate attention to recipient countries instead of fixing domestic problems in their own countries.

4.1.4. Undermining the community conservation strategy

There is a growing consensus globally that provision of tangible economic benefits to communities bordering the protected areas is the right strategy towards minimizing human-wildlife conflicts and motivating the communities to align their bahaviours with conservation goals by refraining from unsustainable behaviours and actions that are destructive to natural resources. The guiding principle to this is that an incentive to conserve, and to tolerate wildlife-related costs, among the local communities is a function of economic gain [See e.g. 35-36]. Under economic recession, where most of the revenues are intended to come from tourism sector and donors, it is unlikely that this strategy can work flawlessly. Since economic benefits are regarded as important condition for behaviour change, it may not be surprising if the communities will resort to illegal activities and, therefore, increase pressures in protected areas. The likelihood of this scenario increases as the local economy also suffers from recession, causing increased incidences of food insecurity and income poverty among the households.

4.1.5. Inefficient state law enforcement

Global economic recession triggers meager budget for natural resources sector and management of the protected areas, in particular. Logically, the situation is worsened by reduced tourism revenues and donor support along with minimal priority accorded by government to conservation compared to other sectors. The underfunding of the protected areas leads to inadequate staffing, inadequate and poor equipment and, consequently, failure to enforce the conservation laws effectively. The 1970s to 1980s global economic recession can epitomize this scenario. The recession rendered the entire natural resources sector (i.e., wildlife, land, forestry, and fisheries) getting only 1.2% of the total national development budget [38]. While the actual cost for effective control of poaching was estimated to range from US$200 to 400/km^2 per annum [16; 39], the budget for big protected areas, such as the Selous Game Reserve, were as low as US$3/km^2 [40]. The staff-area ratio in most protected areas were 1:125 (persons:km^2), far below the recommended ideal ratio of 1:25 [41].

The underfunding of the sector caused huge loss of the populations of two of Africa's charismatic species – rhino and elephant. In 1976, for example, an aerial census estimated 110,000 elephants in the Selous Game Reserve in Southern Tanzania. Uncontrolled poaching reduced this population by 50% in 1986 and to approximately 22,000 in 1991. The rhino population in the reserve dropped from 2,500 in 1976 to 50 in 1986 and zero in 1991 [42]. Similarly, poaching in Serengeti National Park drove the black rhino to the verge of extinction while the elephant population dropped by 80% [43]. Countrywide, the elephant population dropped from 306,300 individuals in 1976 [42] to 203,900 individuals in 1981, to 100,000 in 1987 and to 57,334 in 1991 [44]. About 275 rhinos remained in 1992 compared to 3,795 individuals in 1981 [45].

4.2. Climate change

Climate change is one of the emerging challenges of the 21st century. Tanzania, like other developing countries, is "highly vulnerable" to the impacts of climate change "because of the factors such as widespread poverty, recurrent droughts, inequitable land distribution, and over-dependence on rain-fed agriculture" [46]. Experts predict the possibility of extreme events posing the greatest climate change threat to Africa [47], including Tanzania, where the frequency, intensity and unpredictability of drought, floods and tropical storms are expected to increase. The wildlife protected areas are not and cannot be exempted from the impacts of climate change. The circumstances through which climate change can negatively affect the protected areas include:

4.2.1. Increasing of illegal activities

Low crop yield and death of livestock among the agricultural communities around the protected areas due to droughts, floods and diseases exacerbate poverty. When such situation happens the poor often resort to pursuing illegal and unsustainable activities inside and around the protected areas. For example, studies in Serengeti National Park have shown that illegal hunting is high among the poor households and increases at bad years when the crop yield is low [27, 28]. Similarly, illegal grazing of livestock inside the protected areas increases during the severe droughts. This is due to reality that unlike unprotected lands, protected areas often contain abundant and higher quality pasture during the drought seasons. The livestock owners, therefore, trespass and graze their livestock illegally inside the protected areas leading to serious conflicts between wildlife staff and local communities. In many protected areas such as Kijereshi and Maswa Game Reserves, these conflicts have culminated into wounding and killing of wildlife staff [48]. Oftentimes pastoralists have coped with droughts by moving with their livestock to other parts of the country where they equally increase pressure in protected areas' exceptional resources and values. For instance, movement of Sukuma pastoralists towards southern Tanzania in 1990s and 2000s had serious ecological impacts in Ihefu and Great Ruaha River, which are key for survival of Ruaha National Park [49]. Experience shows that in many protected areas, illegal activities such as poaching increase when events such as floods destroy the infrastructure and making the parts of the protected areas inaccessible by law enforcement staff (personal experience).

Figure 3. The impacts of climate change like this compel livestock owners to graze their livestock inside the protected areas illegally.

4.2.2. Increase of the incidences of wild fires

Incidences of fire become more severe during the extreme droughts and, thus, killing wildlife species, destroying forage resources, reducing water supply and habitats. A study by Hemp [50] showed that loss of forest cover as a result of fire intensity and forest clearing in Kilimanjaro National Park has a more devastating impact than the melting glaciers. According to author glacier contributes one million cubic meters to water supply, while forest cover contributes 500 million cubic metres. Forest and bush fires have also contributed to the destruction of forest resources in the Uluguru Mountains Nature Reserve, which could have similar implications for the water security of downstream communities [47].

Figure 4. Increased poverty leaves the poor without option other than poaching. Four suspected poachers arrested in Ngorongoro Conservation Area with poisoned watermelons and pumpkins targeted to kill the elephants.

4.2.3. Impact on tourism industry

The floods and other climatic hazards affect the infrastructure such as roads and, therefore, render the protected areas, which are key tourist destinations, inaccessible. These consequently, reduce revenues which are important sources of funds for conservation work. A good example is the 1997/98 *El Niño* episode, which rendered most of the areas in the Tanzania's northern tourist circuit inaccessible. In order to cope with poor and inaccessible roads attributed to heavy rains in Serengeti and Arusha National Parks, various local tour operators resorted to taking their visitors around the park using tractors. The farming machines were used as path-finders or to perform the task of dragging, pulling or jostling tour vehicles that were stuck in the rain drenched, soggy grounds of the parks [51]. The heavy downpours also caused several airstrips in the parks, including the most important, Seronera to be closed down.

4.2.4. Increased human-wildlife conflicts

Human-wildlife conflicts often increase during the extreme droughts. This is the time when illegal grazing of livestock occurs inside the protected areas as pasture becomes scarce in Illegal livestock grazing is a serious management issue in Maswa, Ibanda, Burigi, Biharamulo, Moyovosi, Ugalla, Kimisi and Kitengule Game Reserves and Tarangire National Park and Kilombero Ramsar Site. Illegal grazing in protected areas is sometimes associated with widespread use of poison against predators in retaliation for livestock depredation. In Ibanda Game Reserve, for instance, this has led to local extinction of lions (Hassan Mnkeni, pers. comm). On the other hand, wild animals move out of from the protected areas and cause crop damage, livestock depredation and accidents to people. These scenarios occur in virtually all protected areas in Tanzania and they jeopardize the integrity of the protected areas.

5.2.5. Increased risk of species extinction

Extreme droughts and floods cause deaths to numerous wildlife species through destruction of important resources such as forage, water and shelter along with increasing incidences of diseases. For example, the aftermath of *El-Nino/La-nina* weather spells, in the Simanjiro District and Ngorongoro Conservation Area were reported to have brought forth the huge swarms of deadly insects known as "Stomoxys" which claimed the lives of both livestock and wildlife by inflicting bad wounds and painful sores to the animals. The first outbreak of Stomoxys flies occurred in 1962 following the extensive drought of 1961, followed by heavy rains of 1962. The epidemic resulted into the death of over 67 lions [52].

The wildlife species which are globally threatened due to factors such as low population numbers, restricted or patchy habitats, limited climatic ranges and/or restricted habitat requirements are more exposed to risk of extinction than others. Based on this reality, the Intergovernmental Panel on Climate Change (IPCC) warns that climate change will worsen the risk to these species if effective mitigation and adaptation measures will not be implemented.

Recent report by UN Food and Agricultural Organization (FAO) indicates that about 200 animal species in Tanzania classified by IUCN as vulnerable, endangered or critically endangered are subjected to more risk due to effects of climate change [53]. Of these species, are the large mammals including charismatic and flagship species such as elephant (vulnerable), black rhino (critically endangered), wild dog (endangered), cheetah (vulnerable), lion and abbott's duiker (vulnerable). These species constitute one of the key exceptional resource values in many Tanzanian protected areas. Therefore, their loss will obviously affect the tourism industry and lower the revenues which are important source of funds needed for conservation work.

4.3. HIV/AIDS pandemic

HIV/AIDS, one of the worst pandemics in history, touches virtually all sectors in Africa including natural resources sector. However, its appreciation in the conservation literature is still minimal. Empirical data to quantify the impacts of the pandemic on the sector are lacking, though the situation is seemingly to be alarming. While, there is a need for scientific studies to quantify the impacts of HIV/AIDS pandemic in Tanzanian protected areas, the link between this pandemic and wildlife management can be explained as follows:

4.3.1. Weakened performance in the protected areas

Increased rates of illnesses and deaths among the protected areas rangers, senior officials, community game guards and other conservation personnel weaken the performance in the protected areas [54, 55]. This is likely to be the case as wildlife staff can hardly execute their duties including law enforcement when they are sick. Even the most committed employees become unproductive since successive bereavement undermines morale and enthusiasm. Poachers may take advantage to hunt illegally when wildlife staff members are sick, looking after their sick relatives or attending funerals. Economically, HIV/AIDS pandemic imposes huge financial costs to government, conservation agencies and communities. The following impacts of HIV/AIDS on conservation organizations in Africa, adopted from UNAIDS [55], are applicable to Tanzania situation and, protected areas in particular.:

- **Loss of investment in training:** Many conservation organizations have lost highly trained staff to the epidemic. This is particularly serious in Africa, where conservation capacity is already limited. Training replacement staff is very expensive – if funds are available at all.
- **Loss of staff time:** There is an increased absence from work when staff members care for their sick family members and attend funerals of relatives, friends and colleagues.

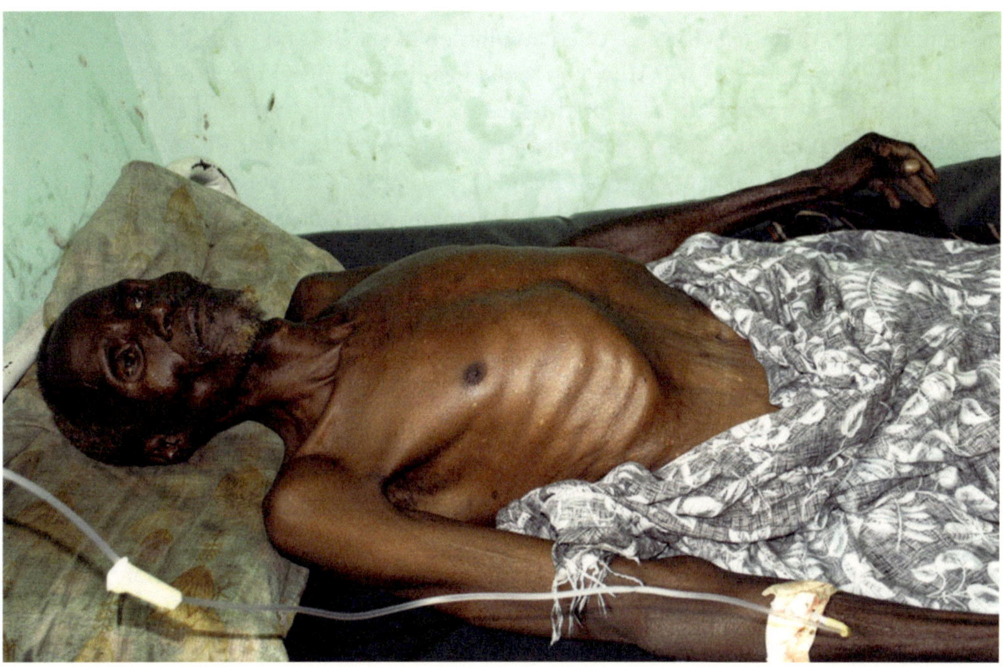

Figure 5. HIV/AIDS and associated opportunistic diseases undermine the performance of wildlife staff and causes overexploitation of natural . in the protected areas

- **Diversion of conservation funds for AIDS costs:** Many conservation organizations are covering the costs of medical expenses, sick leaves, terminal benefits, funeral costs, and training for replacement staff. These expenses reduce the budget available for conservation work, and often have to be covered by scarce core funds.
- **Decline in morale:** Successive bereavement saps morale and enthusiasm from even the most committed employees, slowing productivity.

4.3.2. Increased illegal activities due to household poverty

Agriculture is the leading employer in Africa and other developing countries. However, the sector is threatened by AIDS-related deaths among farm workers, most notably in southern and eastern Africa [56]. UN Food and Agricultural Organization (FAO) projected that 16 million agricultural workers would die of AIDS in 25 African countries with high rates of HIV prevalence between 2000 and 2020 [57]. Low agricultural production and food insecurity translates into increased poverty among the local communities. Households which have lost their breadwinners through HIV/AIDS pandemic remain with no option for meeting their subsistence and income needs. They, therefore, switch to coping strategies that are unsustainable and ecologically destructive such as killing of wildlife species and clearing of habitats.

4.3.3. Overexploitation of natural resources for medicinal use

Available research-based literature indicates that HIV/AIDS had had some serious environmental implications through overexploitation of species and habitat destruction [55]. In Tanzania, prevalence of HIV/AIDS pandemic has roused beliefs that have contributed to these problems. A number of traditional healers are capitalizing on pandemic by claiming that they can treat the pandemic and related opportunistic and chronic diseases that western trained doctors cannot. For instance, in the past, poaching of giraffe was not an issue that could draw considerable conservation or management attention among the protected area managers. However, of recent it is becoming a major issue following a belief among the people that brain and bone-marrow from this species can cure HIV/AIDS. In the period between 2004 and 2008, mass poaching of giraffes was reported in Monduli District and the West Kilimanjaro Wildlife Corridor - striding between Arusha and Kilimanjaro National Parks [58, 59]. In 2011, a retired pastor in Samunge Village of Loliondo Division, Ambilikile Mwasapile, claimed that he was ordered by God through a dream to dispense the herb, *Carisa edulis,* to heal the sick suffering from all chronic diseases including AIDS, diabetes, and asthma. The publicized news about miracle cure caused an influx of thousands of people from all over East Africa. A cup (or *Kikombe*) of the herbal concoction was regarded as a sufficient dose for all diseases. Serengeti and Ngorongoro Conservation Area and Loliondo Game Controlled Area, which are close to the village suffered through habitat destruction (deforestation for firewood and physical impacts of vehicles) and pollution (human wastes and garbage) as roads to Samunge village pass through these protected areas (Figure 6).

Figure 6. Vehicles going to Samunge village and people queuing for herbal concoction from a retired Lutheran Pastor, Ambilikile Mwasapile, who claimed to have received revelation of medicine from God through a dream that can cure all chronic diseases including AIDS. The queues of vehicles with patients who were waiting to drink the concoction reached up to 46-kilometre long. On average, over 4000 patients were served per day.

4.4. Political instability/civil wars

Political instability - defined as the unsteadiness in governments, regime changes and the insecurity that the society receives out of these changes in a nation or in a region - is endemic to many African countries. The causes of political conflicts and instability in Africa have political, economic and social-cultural dimensions [60, 61]. Political causes of conflicts and instability include the struggle for power; lack of visionary leadership; external influence; lack of good governance and transparency; and abuse of human rights. Economic causes include a deterioration and deep malaise of the economy, widespread poverty and a large pool of unemployed, landless and aimless youth; inequitable distribution of resources and national wealth and the negative effect of the external debt burden and the international financial system. Social and cultural causes include social inequality; system of exclusion and ethnic hatred; role of the political class in the manipulation of ethnic and regional sentiments; cultural detachment and the search for identity with extra-African culture; and defective educational system [60, 61]. While some causes of instability are purely internal and portray specific sub-regional dynamics, others have a significant international dimension [61]. International interests have often been a cause of conflicts for political and economic reasons. As a result many countries endowed with abundant natural resources are subjected to higher risk of civil wars making these countries' resources a curse instead of being a blessing.

4.4.1. Reduced revenues from tourism sector

Political instability is bad news for a country's tourism industry, even if no tourist ever becomes physically harmed or killed. This is due to natural sensitivity of tourists to events of political instability and violence in their holiday destinations. Political instability and violence jeopardize a relaxed and unconcerned holiday [62]. Political violence forces the tourists to choose an alternative destination with similar characteristics but in a more stable condition. Official authorities in the countries where tourists originate often issue an advice to their citizens against traveling to destinations characterized by the widespread and prolonged violence. Since tourism is the major source of funding of the conservation activities in the protected areas it is apparent that these activities will be affected once the country or its neighbours get into political turmoil. Examples from Tanzania and other countries in East Africa corroborate this reality. For example, bombing of American embassies in Dar es Salaam and Nairobi in 1998 affected tour business and caused a drastic drop of inquiries about holidaying in Tanzania with some potential customers who had already booked for safaris cancelling their bookings [25]. In Kenya, tourism industry suffered 90% drop in arrivals following the 2007 Post Election Violence [63]. Following its land reform programme, western countries labeled Zimbabwe as a dangerous place for tourist to visit. This negative image imposed on Zimbabwe reduced the tourism revenues notably from US$700 million in 1999 to US$71 million in 2003. As a result, over 80% of the country's large game in private conservancies was illegally hunted [64].

4.4.2. Increased poverty and divergence of government priority to strengthen military activities

It is irrefutable that neither individuals nor government agencies and other potential stakeholders can competently concentrate in planning and executing conservation programmes in an environment of war and political turmoil. Furthermore, economic activities can hardly proceed harmoniously in this environment. It is, therefore, likely that most of the people around the protected areas are subjected to hunger and poverty, a scenario which may force them to engage in poaching of wildlife resources from the protected areas. This problem may be simplified by the fact that during the war, law enforcement cannot be conducted efficiently. Experience has also shown that, governments' priority shifts to political crises, leaving other sectors including conservation unsupported. In some countries such as Rwanda, Uganda, DRC, Mozambique and Southern Sudan, protected areas and wildlife species have been used to support the soldiers through provision of shelter and bush meat. In such situation it becomes very difficult to manage the protected areas.

4.4.3. Human population growth

Civil wars are a major population push factor from areas where wars are waged to areas where peace and tranquility prevail. Tanzania, unlike its neighbours had never experienced the civil wars but the impacts of these wars had been felt in its protected areas and, conservation sector in general. Civil wars and political instability contribute to population growth through influx of refugees. For example, political instability in Rwanda, Burundi and the Democratic Republic of Congo in 1990s caused an influx of more than a million refugees at one time. This had far-reaching effects by causing overexploitation of natural resources and environmental degradation in and around the protected areas located in the western part of the country (including Burigi, Biharamulo, Ibanda and Rumanyika Game Reserves) as expounded below:

4.4.4. Illegal hunting

The prolonged presence of refugees in western Tanzania and possession of sophisticated firearms caused rampant poaching of wildlife species for meat [65 - 69]. Essentially, demand for wild meat has been driven partly by insufficient refugee food rations that failed to supply meat protein [69]. An average number of wild animals which were killed from the game reserves every day to supply animal protein were estimated at 100 [65]. Statistics indicate that majority of the arrested poachers were refugees. In Kagera Region, 87% of arrested poachers in the mid-1990s were refugees [69]. In Ibanda and Rumanyika Game Reserves, refugees arrested as poachers exceeded 60% [65]. Proximity to Great BENACO Refugee Camp made Burigi Game Reserve suffer most. Over 3,000 poachers were arrested in a year period in this Reserve. These illegal activities associated with refugees resulted to a dramatic decline of wildlife species. For, example, animal census conducted by Tanzania Wildlife Conservation Monitoring (TWCM) in Burigi-Biharamulo Game Reserves in 1990 and 1998 indicated that the reserves had lost about 90% of the populations of 13 ungulates (Table 2).

s/n	Animal species	1990 Estimates	1998 Estimates	% loss
1	Bushbuck (*Tragelaphus scriptus*)	229	18	92
2	Eland (*Tragelaphus oryx*)	878	237	73
4	Impala *Aepyceros melampus*)	5,130	2,795	56
5	Lichtenstein's Hartebeest (*Alcelaphus lichtensteini*)	324	0	100
6	Reedbuck (*Redunca redunca*)	147	98	33
7	Roan Antelope (*Hippotragus equines*)	466	15	97
8	Sable Antelope (*Hippotragus niger*)	279	32	89
9	Sitatunga (*Tragelaphus spekei*)	490	0	100
10	Topi (*Damaliscus korrigum*)	6,399	160	97
11	Waterbuck (*Kobus ellipsiprymnus*)	822	94	89
12	Warthog (*Phacochaerus aethiopicus*)	2,628	71	97
13	Zebra (*Equus burchelli*)	6,552	606	91

Source: TWCM [70, 71].

Table 2. Comparison of 1990 and 1998 wet season estimates for common wildlife species in Burigi-Biharamulo Game Reserves

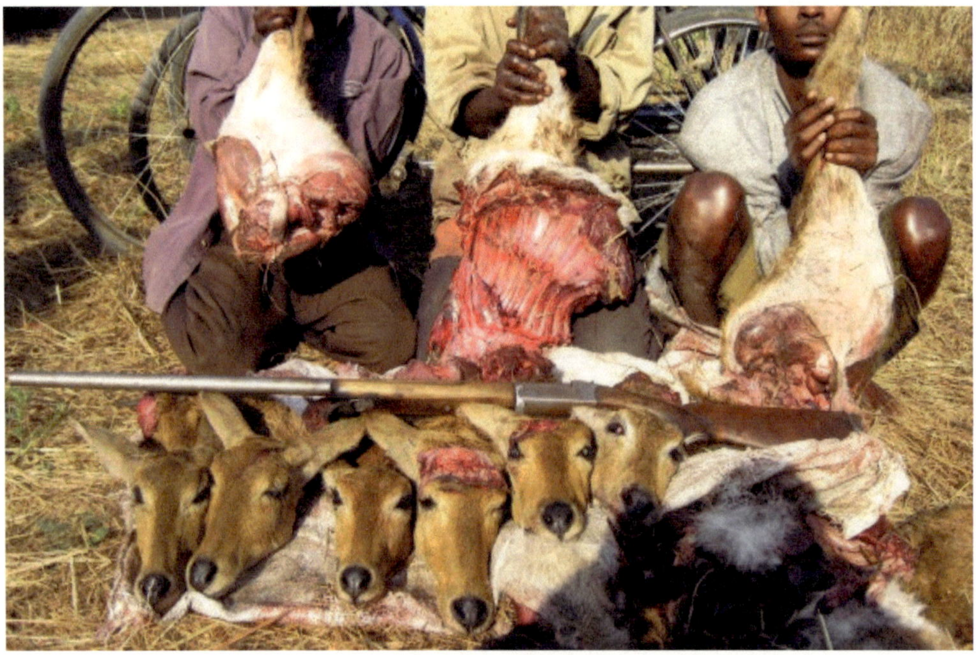

Figure 7. Illegal hunting for bush meat is important coping strategy against poverty

The impacts of refugees were also noted in Gombe National Park. Numbers of several wildlife species including buffalo, zebra, bushbuck, and duiker (*Cephalophus* spp.) were reported to have declined notably [69]. Also noted in southern portion of this park was a considerable deccrease of the population of chimpanzee (*Pan Troglodyte*) attributed to proximity of the area with large Congolese immigrants, who traditionally eat primate meat [69].

4.4.5. Habitat destruction

Along with illegal hunting, refugees had a profound impact on wildlife habitats. Deforestation caused scarcity of fuel resources, land degradation, destruction of water sources and, consequently, encroachment into protected areas. At the peak of the Rwanda refugee crisis, daily consumption of firewood for camps in the Kagera region alone was about 1,200 tons [66]. Generally, an average of 300 metric tons of fuel wood were consumed per day in 1997 [65]. The impacts of deforestation extended up to 20km away from the camps. Destruction or deforestation in BENACO area was estimated at 960 km^2 of land. Aerial photos of the affected region taken in 1996 showed that some 225km^2 and roughly 470km^2 of land were completely and partially deforested, respectively [65].

Figure 8. Refugees fleeing civil wars from their countries contribute to population increase and demand for resources at the destinations where they settle.

5. Conclusion and the way forward

The reviews presented in this chapter provide unquestionable reality that global economic recession, climate change, HIV/AIDS pandemic and political instability are potential factors, among many others, that undermine the efforts geared towards the management of the protected areas. There is direct and indirect links between these issues and loss of wildlife habitats and species in many protected areas. It is, therefore, imperative that these issues are accorded adequate priority by mainstreaming them into policies and management plans of the protected areas and conservation agencies. The effective strategies for addressing these issues should be developed and form a part of management plans for protected areas. The following are some specific recommendations for each of the issues.

5.1. Economic recession

The financing of protected areas in Tanzania heavily relies on international tourists and donors. However, as shown earlier, these sources are vulnerable to a number of factors including global economic recessions. Unfortunately, Tanzania lacks preparedness mechanisms to offset the effects of economic recessions in protected areas. This deficiency should be addressed. The possible approach is to establish the sustainable financing mechanisms that will guarantee the continued existence and integrity of the country's protected areas. The following actions adopted from Runyoro and Kideghesho [25] are recommended:

- Development of the "Conservation Trust Fund". Trust funds have been established in many developing countries over the past decade as a way of providing long-term funding for protected areas. Trust funds are typically legally independent institutions managed by independent boards of directors and have a permanent endowment that is supported through grants.
- Tanzania should be promoted together with other East African Community countries as one tourism destination and an elaborate and sustainable tourism for domestic, regional and African Continent citizens should be promoted and encouraged to visit Tanzania's attractions more frequently as much as the government commits itself to improving infrastructure and services along with mainitaining peace and tranguility.
- The development of a revenue retention scheme similar to that of Selous Game Reserve that would increase the local capacity of the conservation agencies to manage the protected areas under their jurisdiction.
- The Government of Tanzania should consider relieving taxing government organizations entrusted to manage the protected areas in order to improve the tourism industry as the act of taxation has become a burden and an impediment to ensuring high class conservation of these resources.

5.2. Climate change

The problem of climate change and its potential impacts on protected areas can be addressed by adoption of a variety of mitigation and adaptation strategies. The possible strategies include:

- The protected area and conservation managers should be familiar and understand the importance and relevance of climate change and adaptation. This may necessitate capacity building through offering training that will equip the managers with relevant skills and knowledge. This will enable them to critically analyze the current exposure to climate shocks and stresses, and provide a model-based analysis of future impacts of the problem. Capacity can be developed through: briefings; training materials; short courses for staff and partners; and regular knowledge and information exchange between staff and partners working in different sectors and in 'lessons learnt'.
- Protected area and conservation managers in collaboration with other stakeholders should work out the strategies for reducing vulnerability to climate change as one of the priority agenda. To this end, the protected area managers, conservation agencies and other stakeholders must focus on building adaptive capacity, particularly to the most vulnerable people; and, in some cases, on reducing exposure or sensitivity to climate

impacts. The precaution should be taken to ensure that development initiatives do not inadvertently increase vulnerability. Effective reduction of vulnerability will reduce much of the pressures in protected areas from the people who would look at protected areas as the only possibility for their survival.

5.3. HIV/AIDS pandemic

The damaging impacts of HIV/AIDS pandemic on conservation sector and protected areas prompts the need to rank this challenge among the top priorities in the management plans of the respective protected areas. The following actions should be observed:

- The protected area managers and conservation agencies should mainstream HIV/AIDS into their policies and management plans. UNAIDS and World Bank [72] define mainstreaming HIV/AIDS as the process that enables the actors to address the causes and effects of HIV/AIDS in an effective and sustained manner, both through their usual work and within their workplace. It means "wearing AIDS glasses" while working in all sectors and at all levels. Essentially, mainstreaming HIV/AIDS means all sectors determining: the ways through which they may contribute to the spread of HIV/AIDS pandemic; the ways in which the epidemic is likely to affect their sector's goals, objectives and programmes and where their sector has a comparative advantage to respond to and limit the spread of HIV and to mitigate the impact of the epidemic [73].
- Ensure that all factors driving the HIV/AIDS epidemic such as poverty and gender inequalities are sufficiently addressed by the management authorities of the protected areas, conservation agencies and the government. This may involve developing policies that address gender equality and human rights along with adopting sustainable poverty reduction strategies that will strengthen people's livelihoods and therefore preempt the need to obtain resources from protected areas illegally and unsustainably.
- Mobilizing the public and private stakeholders to actively take part in the implementation of strategies aiming at fighting the epidemic in and around the protected areas. The strategies, among others, should include promotion of high level advocacy and education on HIV/AIDS pandemic, protection of human and communal rights of people infected and affected with HIV/AIDS, enhancing health care and counseling of HIV/AIDS patients, ensuring the welfare of the bereaved orphans and survivors of HIV/AIDS victims and handling of social, economic, cultural and legal issues related to this epidemic.

5.4. Political instability

Detrimental impacts caused by civil wars in protected areas through degradation and loss of biodiversity, calls for adoption of a number of strategies –those required to prevent occurrence of conflicts and political instability as well as those required to mitigate the problems and impacts caused by these situations(in case they occur). The following are possible strategies:

- Strategies for conflict prevention and peace building should be sought. One way towards this end is to ensure that the principles of good governance and accountability are observed by all countries and all sectors. International community, when necessary, should intervene to fight social vices which can lead to civil wars such as inequalities, injustice, corruption, nepotism etc. Furthermore, in order to ensure that peace and tranquility are sustained for longer time there is a need for establishment of a global network on conflict prevention and peace education in collaboration with relevant ministries and organizations in several countries. Civil societies and religious organizations, among others, should take a lead to this end.

- The conservation community should view the problem of refugees, not only as political, but also as ecological challenge. Therefore, there is a need for conservation authorities to collaborate with other stakeholders to ensure that the ecological problems brought by refugees in protected and adjacent areas are addressed.
- Conservation managers should assume a new role as advocates of peace at local, regional and global levels. It is true that historically the impacts of political instability in Tanzania have been felt in the protected areas located in the periphery regions as the problem has often being emanating from the neighbouring countries. This is due to fact that, for years, Tanzania has enjoyed peace and tranquility and, therefore, internal political environment had rarely seemed to affect the management of protected areas. However, this scenario should not be considered as a prerogative to Tanzania. The fact that the political climate and socio-economic and ecological factors are changing may change the situation to worse if pragmatic measures will not be taken to cope and adapt to these changes.
- The international community should ensure that all factors driving the refugees to behave unsustainably by poaching and destroying habitats are adequately addressed. These entail provision of adequate food and alternative fuel for cooking and heating.
- When the problem of refugees arises, the government and other stakeholders should work out the logistics to distribute the refugees to different parts of the country in order to minimize pressure on resources and habitats caused by concentration of refugees in one place.

Author details

Jafari R. Kideghesho
Sokoine University of Agriculture (SUA), Tanzania

Tuli S. Msuya
Tanzania Forestry Research Institute (TAFORI), Tanzania

6. References

[1] IUCN. Guidelines for Protected Area Management Categories. Gland and Cambridge: IUCN; 1994.

[2] Lopoukhine N. Protected areas — For Life's Sake. In: Secretariat of the Convention on Biological Diversity (Ed). Protected Areas in Today's World: Their Values and Benefits for the Welfare of the Planet. Montreal: Technical Series; 2008 p. 1-3.

[3] Scherl L.M. and Emerton L. Protected areas and Poverty Reduction. In: Secretariat of the Convention on Biological Diversity (Ed). Protected Areas in Today's World: Their Values and Benefits for the Welfare of the Planet. Montreal: Technical Series; p 4-18.

[4] WWF International. Guidelines for Community-based Ecotourism Development. Gland: WWF International;2001.

[5] Mitchell J, Ashley C. Can Tourism Reduce Poverty in Africa? ODI Briefing Paper. Overseas Development Institute 2006; www.odi.org.uk/publications/briefing/bp_March06_Tourism1-web.pdf. (accessed 10 May 2012).

[6] Okayasu S. The Status of Ecotourism in the Eastern Arc Mountains of Tanzania. MSc. Thesis, Imperial College London; 2008.

[7] UNWTO (2007) Tourism Highlights, Edition 2007. UNWTO Tourism Highlights. United Nations World Tourism Organization.

[8] IUCN (World Conservation Union). 1996 Redlist of Threatened Species. IUCN;1996. www.iucnredlist.org. (accessed 20 March 2012)

[9] IUCN (World Conservation Union). 2008 Redlist of Threatened Species. IUCN;2008. www.iucnredlist.org. (accessed 20 March 2012)

[10] Cadman, M. Poachers target elephants, rhinos. 2007;Independent Online. [Online]. Available: http://www.iol.co.za/in (accessed 20 October 2011)

[11] UNEP. World Database on Protected Areas Annual Release. Cambridge:UNEP-WCMC; 2010. http://www.wdpa.org/AnnualRelease.aspx#expandText4 (accessed 10 June 2012).

[12] CBD. 2010 Biodiversity Target. CBD; 2010. http://www.cbd.int/2010-target/. (accessed 10 June 2012).

[13] UNEP-WCMC. World Database on Protected Areas. Cambridge:. UNEP-WCMC;2009. http://www.wdpa.org/. (accessed June 10, 2012).

[14] Mora C, Sale P. Ongoing Global Biodiversity Loss and the Need to Move Beyond Protected Areas: A Review of the Technical and Practical Shortcoming of Protected Areas on Land and Sea. Marine Ecology Progress Series. 2011;(434):251-266.

[15] URT. Wildlife Policy of Tanzania. Dar es Salaam:Government Printer;1998.

[16] Bonner R. At the Hand of Man: Peril and Hope for Africa's Wildlife. New York:Alfred A. Knopf;1993.

[17] Neumann RP. Political Ecology of Wildlife Conservation in the Mt Meru Area of Northern Tanzania. Land Degradation and Rehabilitation 1992;3:99 – 113

[18] Nash RF. Wilderness and the American Mind. NH:Yale University Press;1982.

[19] UNEP-WCMC. World Database on Protected Areas. Cambridge:. UNEP-WCMC;2005. http://www.wdpa.org/. (accessed June 10, 2012).

[20] URT. Wildlife Policy of Tanzania. Dar es Salaam:Government Printer;2007.

[21] Kideghesho JR, RunyoroVA, Nyahongo JW, Kaswamila A. 50 years of Arusha Manifesto and the future of wildlife conservation in Tanzania. 2011: Conference Proceedings December 6-8, 2011, Arusha International Conference Centre, Arusha, Tanzania. Arusha:Tanzania Wildlife Research Institute (TAWIRI);2011.

[22] World Bank. Global Development Finance 2008. Washington DC: World Bank;2008.

[23] Massa I, te Velde WD. The Global Financial Crisis: Will Successful African Countries be Affected?', ODI Background Note, 8 December. London: Overseas Development Institute;2008.

[24] Lunogelo HB, Mbilinyi A, Hangi M. The Global Financial Crisis and Tanzania: Effects and Policy Responses. Final Report submitted to the Economic and Social Research Foundation (ESRF). Dar es Salaam: ESRF 2009.

[25] Runyoro VA, Kideghesho JR. Coping With the Effects of Global Economic Recession on Tourism Industry in Tanzania: Are We Prepared? Journal of Tourism Challenges and Trends 2010; 3(1)95-110.

[26] Ngowi HP. The Current Global Economic Crisis and Its Impacts in Tanzania. African Journal of Business Management 2010;4(8):1468-1474

[27] Kideghesho, JR, Røskaft, E, Kaltenborn, BP, Tarimo, TCM. Serengeti shall not die: Can the ambition be sustained? The International Journal of Biodiversity Science and Management 2005;1(3)150-166

[28] Loibooki M, Hofer H, Campbell KLI, East ML. Bushmeat Hunting by Communities Adjacent to the Serengeti National Park, Tanzania: The Importance of Livestock Ownership and Alternative Sources of Protein and Income. Environmental Conservation 2002; 29:391–398.

[29] Thirgood S., Mlingwa C, Gereta E, Runyoro VA, Malpas R, Laurenson K, Borner M. Who Pays for Conservation? Current and Future Financing Scenarios for the Serengeti Ecosystem. In: Sinclair ARE, Packer C, Mduma SAR, Fryxell JM (Eds.). Serengeti III. Human Impacts on Ecosystem Dynamics. Chicago:University of Chicago Press; 2008. p. 443-469.

[30] TANAPA. TANAPA TODAY: A quarterly Publication of Tanzania National Parks. Arusha: TANAPA;2009.

[31] NCAA. (2009). Mid-Term Expenditure Framework (MTEF) Budget 2009/2010. Arusha; Ngorongoro Conservation Area 2009.

[32] Emerton L., Bishop J, Thomas L. Sustainable Financing of Protected Areas: A Global Review of Challenges and Options. Gland and Cambridge: IUCN; 2006.

[33] Hance, J. Economic Crisis Threatens Conservation Programs and Endangered Species. An Interview with Paula Kahumbu of WildlifeDirect on August 17, 2009. http://news.mongabay.com/2009/0817-hance_kahumbu.html. (accessed on 12 June 2012).

[34] Butler, R. Economic Crisis Hits Conservation but May Offer Opportunities: An Interview with Mark Tercek, the President and CEO of the Nature Conservancy, One of the World's Largest Conservation Group on March 03, 2009. http://news.mongabay.com/2009/0303-tercek_interview.html. (accessed on 12 June 2012.

[35] Little PD. The Link Between Local Participation and Improved Conservation: A Review of Issues and Experiences. In: Western D., Wright, M. and Strum, S. (Eds). Natural Connections: Perspectives in Community Based Conservation. Washington DC: Island Press;1994. p. 347-372.

[36] Wells M, Brandon K. People and Parks: Linking Protected Area Management with Local Communities. Washington DC: World Bank; 1992.

[37] Western D. The Background to Community-based Conservation. In: Western D., Wright, M. and Strum, S. (Eds). Natural Connections: Perspectives in Community Based Conservation. Washington DC: Island Press;1994. p. 1-12.

[38] Yeager R. Land Use and Wildlife in Modern Tanzania. In: R. Yeager, Miller NN (Eds.). Wildlife, Wild Death: Land Use and Survival in Eastern Africa. New York: State University of New York Press; 1986. p. 21-65.

[39] Leader-Williams N, Albon SD, Berry PSM.. Illegal Exploitationof Black Rhinoceros and Elephant Populations: Patterns of Decline, Law Enforcement and Patrol Efforts in Luangwa Valley,Zambia. Journal of Applied Ecology 1990;27:1055–1087.

[40] Baldus R, Kibonde B, Siege L. Seeking conservation partnership in the Selous Game Reserve, Tanzania. Parks 2003;13:50–61.

[41] URT. A Review of the Wildlife Sector in Tanzania. Volume I: Assessment of the Current Situation. Dar es Salaam: Wildlife Sector Review Task Force; 2005

[42] Nelson F, Nshala R, Rogers WA. The Evolution and Reform of Tanzanian Wildlife Management. Conservation and Society 2007;5:232-261

[43] Dublin HT, Douglas-Hamilton I. Status and Trends of Elephants in the Serengeti-Mara Ecosystem. African Journal of Ecology 1987;25:19-33.

[44] Cumming DHM, Du Toit, RF and Stuart SN. African Elephants and Rhinos: Status Survey and Conservation Action Plan Volume 10 of IUCN/SSC Action Plans for the Conservation of Biological Diversity. Gland: IUCN;1990.

[45] Sas-Rolfes M. Environment briefing paper: Does CITES Work? Four Case Studies. London: Institute of Economic Affairs; 1997.

[46] IPCC (Intergovernmental Panel on Climate Change) . Climate Change 2001: Synthesis Report Summary for Policy Makers. Third Assessment. http://www.ipcc.ch/pdf/climate-changes-2001/synthesis-spm/synthesis-spm-en.pdf. Wembley: IPCC;2001. (accessed 22 June 2012)

[47] Ehrhart C, Twena M. Climate Change and Poverty in Tanzania: Realities and Response Options for CARE. Background Report to CARE International Poverty-Climate Change Initiative. Dar es Salaam: CARE;2006.

[48] Kideghesho JR. Wildlife Conservation and Local Land Use Conflicts in Western Serengeti Corridor, Tanzania. PhD Thesis, Norwegian University of Science and Technology;2006.

[49] Kashaigili JJ, Rajabu K, Masolwa P. Freshwater Management and Climate Change Adaptation: Experiences from the Great Ruaha River Catchment in Tanzania. Climate and Development 2009;1(3): 220-228.

[50] Hemp A. Climate Change-Driven Forest Fires Marginalise the Impact of Ice Cap Wasting on Kilimanjaro. Global Change Biology 2005;11:1013-1023.

[51] Nkwame VM. Tractors Used to 'Game-Drive' Tourists in Serengeti: Tour Operators Want TANAPA to Take Responsibility on Poor Park Roads.
www.arushatimes.co.tz/2007/3/Tourism_1.htm (accessed 30 June 2012).

[52] Arusha Times. Deadly Insects Plagued Crater. http://www.ntz.info/gen/n00103.html. (accessed 30 June 2012).

[53] Mijingo, H. Effect of Climate Change on Tanzanian Wild Species. 2011; http://www.dailynews.co.tz/feature/?n=26459&cat=feature (accessed 18 May 2012).

[54] Ngoti PM, Baldus RD. HIV/AIDS and the Wildlife Sector In Tanzania. Tanzania Wildlife Discussion Paper No.38. Dar Es Salaam: Wildlife Division;2004.

[55] UNAIDS. The Impact of AIDS on People and Societies. 2006 Report on the Global Aids Epidemic. UNAIDS;2006.

[56] De Souza R, Williams JS, Meyerson FAB. Critical Links: Population, Health, and the Environment. Population Bulletin 2003;58(3) 17-28

[57] FAO. "AIDS—A Threat to Rural Africa." FAO (Food and Agriculture Organization) Fact Sheet" www.fao.org/Focus/E/aids6-e.htm, (accessed June 12, 2012).
[58] The Arusha Times. National Pride in Jeopardy. *The Arusha Times.* 4-10, September, 2004. www.arushatimes.co.tz/2010/1/front_page_1.htm, (accessed 21 April 2012).
[59] The Arusha Times. Tanzania: Tale of Momella's Giraffes without Tails. www.arushatimes.co.tz/2010/1/front_page_1.htm, (Accessed 21April 2012).
[60] Bujra A. African Conflicts: Their Causes and Their Political and Social Environment. *DPMF Occasional Paper,* No. 4. Development Policy Management Forum (DPMF). Addis Ababa:United Nations Economic Commission for Africa (UNECA);2002.
[61] Adebayo A., editor. Comprehending and Mastering African Conflicts. London: Zed Books;1999.
[62] Neumayer E. The Impact of Political Violence on Tourism: Dynamic Cross-National Estimation. Journal of Conflict Resolution 2004; 48(2):259-281
[63] Travel News. Kenya's Tourism Industry Suffers 90 per cent Drop in Arrivals. //www.monstersandcritics.com/lifestyle/travel/news/article_1390983.php/Kenyas_tourism_industry (accessed 28 May 2012).
[64] ZimConservation. Report no. 1. 2004 http://zimconservation.com/?page_id=133 (accessed 8 June 2008).
[65] URT. Keynote Address by Honourable Mizengo Peter Pinda (MP), Prime Minister of the United Republic of Tanzania at the Second United Nations High Commissioner for Refugees' Dialogue on Protection Challenges on the Theme of Protracted Refugee Situations in Geneva, December 10, 2008. Dar es Salaam: URT, 2008
[66] UNHCR. Environment in UNHCR. Refugee affected areas. UNHCR Engineering and Environmental Services Section. Switzerland. http://www.unhcr.org/protect/PROTECTION/3b039f3c4.html. (accessed on 3 June 2012).
[67] Jambiya G. Milledge SAH and Mtango, N. *'Night Time Spinach'*: Conservation and Livelihood Implications of Wild Meat Use in Refugee Situations In North-Western Tanzania. Dar es Salaam: TRAFFIC East/Southern Africa;2007.
[68] Berry L. The Impact of Environmental Degradation on Refugee-Host Relations: A Case Study from Tanzania. Switzerland: Policy Development and Evaluation Service, United Nations High Commissioner for Refugees:2008.
[69] Jambiya G., Milledge SAH., Mtango, N, Hurst A. Wild Meat and Food Security in Refugee Hosting Areas. Dar es Salaam: TRAFFIC East/Southern Africa;2008.
[70] TWCM (Tanzania Wildlifer Conservation and Monitoring). Wildlife Census Burigi Biharamulo – 1990. Arusha: TAWIRI;1991.
[71] TWCM (Tanzania Wildlifer Conservation and Monitoring). Wildlife Census Burigi Biharamulo – 1998. Arusha: TAWIRI;1998.
[72] UNAIDS, World Bank. AIDS, Poverty Reduction and Debt Relief: aToolkit for Mainstreaming HIV/AIDS Programmes into Development Instruments. Geneva: UNAIDS / WORLD BANK; 2001.
[73] Elsey H, Kutengule P. Mainstreaming HIV/AIDS in Development Sectors: A Resource Pack for Practitioners and Policy Makers. DFID's HIV/AIDS Knowledge Programme, Liverpool School of Tropical Medicine and Health Economics and HIV/AIDS Research Division. Natal: University of Natal;2003.

Chapter 4

Tara National Park – Resources, Management and Tourist Perception

Jelena Tomićević, Ivana Bjedov, Ivana Gudurić,
Dragica Obratov-Petković and Margaret A. Shannon

Additional information is available at the end of the chapter

http://dx.doi.org/10.5772/51197

1. Introduction

According to an ancient Slavic legend, Tar, king of the gods, chose Tara Mountain with its outstanding and unique natural beauty as the place to spend his divine life [1].

Tara National Park (TNP) was proclaimed a protected natural resource area in 1981 by the First Regulation on the National Park (Official Gazette of RS no. 41/81). According to the Regulation on National Parks of Serbia (Official Gazette of RS no. 39/93), a public enterprise, 'National Park Tara', was founded, with full responsibility for the management of the park [2]. Tara National Park is situated in west Serbia (Figure 1.), the coordinates, according to Greenwich lie between 43°51′ and 43°57′ north, and 17°03′ and 17°11′ east [3]. The region which includes Tara NP extends over an area of 19175 ha. It contains most of Tara Mountain and the region bordered by the elbow-shaped course of the River Drina, between Višegrad and Bajina Bašta, thus belonging to a part of Starovlaške mountains (Starovlaška-Raška Visija highlands) [4].

Tara National Park incorporates the region belonging to the Bajina Bašta municipality. Two local communities, namely Jagoštica and Rastište are situated entirely on the national park territory, with eight further communities partly within the park's boundaries (Perućac, Beserovina, Zaovine, Rača, Mala Reka, Solotuša, Zaugline and Konjska Reka) [5,6]. Five great mountains – Tara, Crni Vrh, Aluške Planine, Zvezda and Kaluđerske Bare – framed by the impressive canyon of the River Drina, represent the park's most precious features. Especially noteworthy is the diversity of the abiogenes and the heterogeneity of the ecological characteristics, as well as a very significant refuge in which numerous relict and endemic species and associations have been preserved, many even since the glaciations. It is considered that certain manmade ecosystems (meadows and pastures) also represent a particular value and potential of this region (Figure 2.) [7].

Figure 1. Map of natural protected areas in the Republic of Serbia, with the geographic position of Tara National Park

Figure 2. Meadows and pastures in Tara NP

Tara mountain range was formed more than 600 million years ago from Palaeozoic limestone and shales. Glacial and postglacial events played a significant part in determining the flora and fauna of this protected area. During the great ice age and the alternation of glacial and interglacial periods, the large Paratetis Sea, part of the Panonian Sea, played an important role. A part of this sea, next to the basin of the River Drina, extended as far as Tara Mountain. Later, the withdrawal and disappearance of the sea caused an alteration of the climate and the formation of specific vegetation [2]. Geomorphologically, the national park region is characterised by a set of mountain humps and highly fissured surfaces bisected by deeply tongued river valleys, with canyon walls of 1000 m in height. The region is made of carbonic, Triassic and chalk rocks [2]. The average altitude is 1000-1200 m. The highest peak in TNP is Kozji rid (1591 m a.s.l.). There are a large number of mountain peaks and ridges, which are excellent for mountaineering and also provide spectacular views. The canyon of the rivers Drina, Derventa and Rača, with their waterfalls and spring Ladjevac, are particularly spectacular [2]. In the national park, agricultural areas cover 3,353 ha, or 17.5 % of total area, comprised of 82 % meadows and pastures, 15 % ploughed fields and 3 % orchards. However, natural resources for the development of agricultural production are limited, because the type of soils differs in their productive value and their capacity to be utilized. There are however, agricultural potentials in terms of meadows and pasture that could support cattle breeding, which have been underdeveloped to date [7].

1.1. Biodiversity of Tara

Thanks to the specific orography of the terrain, the Tara range became a refuge for many species of flora and fauna. This diversity is evident not only in the presence of living species, but also by a great number of relict and endemic species. The most important relict is Panchich's spruce (*Picea omorika*), endemic to this region and a source of national pride, but also an object of interest for world experts, lovers of nature, mountaineers and tourists [8].

The main value of the area is the abundance and biodiversity of natural values. The rich flora is the result of many factors: geographic position, geology, soils, climate, history and altitude[8].

The vascular flora of Serbia contains 3662 taxa [9], of which 1000 plant species have been identified in this region, or one third of the total flora of Serbia, makes TNP the most important area for preservation of biodiversity [4]. According to the research in 1989 [10], the flora of Tara is composed of 35 forest and 9 meadow associations. The majority of species belong to the families Asteraceae, Poaceae, Fabaceae and Lamiaceae. The Asteraceae family accounts for the greatest number of floral species from northern Europe to the Mediterranean [11]. The great number of genera from this family indicates that the Serbian territory is one of the significant development centres of taxonomic differentiation [9].

Forest ecosystems of Tara Mt. are among most diverse and most preserved in Europe. At the lowest elevations the forests are characterised by grey elder (*Alnus incana*), willow (*Salix spp*), European walnut (*Juglans regia*) and the flowering ash (*Fraxinus ornus*). These are succeeded by forests with Austrian oak (*Quercus cerris*), Hungarian oak (*Quercus frainetto*), sessile oak (*Quercus petraea*), Balkan beech (*Fagus moesiaca*), and Austrian and Scot's pine (*Pinus nigra, P. sylvestris*). At the highest elevations theforests consist of silver fir (*Abies alba*), Norway spruce (*Picea abies*) and beech (*Fagus sylvatica*), along with sycamore (*Acer pseudoplatanus*), mountain elm (*Ulmus glabra*), European aspen (*Populus tremula*), etc. A set of impoverished forest associations derived from the Omorikae-Pineto-Abieto-Fagetum mixtum association are found in the national park [2]. Furthermore, there are the plenitude of natural rarities protected by law, such as Panchich's spruce (*Picea omorika*), the hazel tree (*Corylus colurna*), European yew (*Taxus baccata*), European holly (*Ilex aquifolium*), the Derventa knapweed (*Centaurea derventana*), alkanet (*Gentiana lutea*), etc. are a further special feature of the region [2].

Over 250 edible and poisonous mushrooms can be found in the meadows and forests of Tara. The most poisonous is death cap (*Amanita phalloides*). The edible mushrooms of a high quality include: king bolete (*Boletus edulis*), yellow morels (*Morchella esculent*), and delicious lactarius (*Lactarius deliciosus*). The rich fauna consists of a large number of rare but scientifically important species, a number of which are already extinct in many parts of Europe, such as chamois (*Rupicapra rupicapra*) and brown bear (*Ursus arctors*), etc.. There are also many game species: wild boar, wolf, fox, rabbit, marten and wild cat. The mountain complex of Tara, one of the most strictly protected natural areas, provides a habitat for many birds, some of which have already been exterminated throughout most of Serbia (Tomićević, 2005). A small number of golden eagles (*Aquila chrysaetos*) are present, and the capercaillie (*Tetrao urogallus*), for example. In total, 53 species of mammal and 82 bird species have been recorded in this region [2].

Meadows cover large areas of the national park, developing as secondary vegetation on the soils of various former forest associations. Thanks to the diversity of habitats, the Tara area provides meadow associations: - *Ranunculo – nardetum stricte; Danthonietum - calycinae; Cariceto – brometum erecti; Rhinantho – cynosuretum cristati;*

Bromo – plantaginetum mediae; Arrhenatheretum - elatioris; Lythro –caricetum paniculatae; Eriophoretum latifoliae; Patasitetum hybridi [7].The meadow associations are maintained and preserved ecosystems with varied and rich vegetation and fauna. There are a great number of herbs of pharmacological importance, but they have as yet been insufficiently studied.

1.2. Population

The inhabitants of this region belong to the Dinaroid anthropological type. They are highlanders infused by permanent migration streams from southern parts (eastern parts of Herzegovina, Montenegro and Stari Vlah), people of the same physiological features, little changed across centuries and generations [12].

The population is Serbian and lives in scattered villages, so-called 'starovlaski', where the houses of one family make up an independent economic unity. These houses are often far away from one another, and therefore a single village with a small number of inhabitants may sprawl across a number of kilometres. Two villages are situated entirely within the borders of the national park. Jagoštica is a village of the small scattered type, and Rastište consists of strewn hamlets and represents the biggest and most scattered village in the Tara region. Jagoštica is the most isolated settlement and only in the last ten years it has been better connected with neighbouring Rastište and the surrounding settlements [12].

In the period 1948-1981, the population of the Tara region decreased to 5000 people, of which 900, or 17.2%, live within the national park. In Jagoštica village there are 53 households and 163 inhabitants. Rastište village has 107 households and 285 inhabitants. There are eight villages along the borders of the national park. Parts of the associated households and their estates are located within the national park, including mainly forests, but also meadows and pastures. The main occupations of the inhabitants of this region are agriculture and forestry (Figure 3.). A small number of inhabitants of the region are employed, mainly in forestry. The possibility of employment in other activities is limited, leading to a population drain, which along with a low birth rate means that the population is in decline. The dwindling population is a consequence of the underdevelopment of the region and the difficult local employment situation, causing the inhabitants to migrate to more developed areas [13,14].

Figure 3. Cattle breeding in Tara NP

1.3. Cultural and historical heritage

The Tara Mountain possesses a rich cultural and historical heritage. During the periods of Roman and Byzantine occupation of the Balkan Peninsula, the Tara region and the canyon of the River Drina, belonging to the Roman province Ilirikym, represented the most northern natural defensive border. However, the region was located at the margins of both the Roman and Byzantine cultural influence. Therefore, there are no significant remains from those periods. When the Slavs arrived (10th century) in this area, they brought new customs and forms of organisation along with them [2].

With the foundation of the Serbian State in Raška, this region became the defence zone for the state's northern border. The remains of mediaeval fortresses can be found in Perućac and Rastište. In the second part of the 13th century, king Dragutin founded the monastery at Rača, in the canyon of the River Rača. The significance and cultural role of this monastery was especially prominent after the fall of Serbia under Turkish control. In the most difficult period of World War II, the oldest written monument of the Serbian nation – *Miroslavljevo jevandjelje* – was kept in this monastery. Upon the liberation of Serbia from Turkish occupation in the mid 19th century, people from Stari Vlah began to move to the Tara region [2].

Today, these ethnographic characteristics of the region represent a special tourism value. Customs concerning slavas, wedding parties and field work are widespread and specific. The local handicrafts were famous for producing small bags, gloves, socks, jumpers, carpets, flasks and saltcellars, but the tradition of the handicrafts faded away, especially with migration of inhabitants from the villages [1].

1.4. Tourism

Various natural values in Tara National Park, namely specific geomorphological units, good climate and unique vegetation, are a basis for the development of appropriate tourism activities [1]. Tara National Park is wonderfully suited for the development of almost all forms of recreational activities and tourism. The national park is a traditional summer and winter resort. Natural beauty, climate and cultural heritage cater for all kinds of tourism, as well as for sport, recreation, hunting, fishing and hiking. Tourism in Tara has not been well-developed yet. Partly, this is a consequence of the poor traffic infrastructure and a lack of awareness of the importance of tourism for the further development of the region [1].

However, opportunities for tourism in the Tara National Park are great and varied (Figure 4.). They are based on excellent natural conditions for ecotourism, variety of tourist facilities, good transport links and accessibility to attractive and rare tourist attractions and sites: canyons, viewpoints, lakes, special reserves and cultural monuments. Such natural and physical conditions in the area of Tara provide an excellent basis for tourism development, primarily for the purposes of recreation and leisure, as well as active forms of sports and recreational tourism: excursions, hiking, hunting, fishing and rural tourism [15].

1.5. Organisation and management of Tara National Park

The national park is managed according to annual and five-year protection programmes developed by the Institute for Protection of Nature of Serbia, and approved by the Ministry of Environment, Mining and Spatial Planning [16]. The aims and tasks of development were set out in this programme, which is based on the Spatial Plan of Tara National Park. The programme of protection and development of Tara National Park is based on a concept of permanent and balanced development, protection and preservation of natural and manmade features, the preservation of biodiversity, along with the moderate and controlled utilisation of resources with the following aims [2]:

Figure 4. Predov krst – tourist centre in Tara NP

1. Preservation, protection and enhancement of the special natural values of the national park and their utilisation for scientific and other research purposes, education, presentation and recreation according to the ecological potential of the national park;
2. Preservation, enhancement and protection of landscapes within the national park, including the flora, fauna, soil, water, air, pastures, meadows, game and fishing, with utilisation based on the principles of spatial capacities;
3. Development of activities in line with the protection and development functions of the national park (forestry, hunting, fishing, tourism, agriculture, traffic, etc.);
4. Preservation, protection and utilisation of immobile cultural values and all cultural and historical attributes for the purposes of science, education, presentation and recreation;
5. Organised multidisciplinary and long-term scientific research into the phenomena within the national park and the education of all categories of local people and sector branches;
6. Directional development of all existing and potential new activities based on the traditions of the national park region and the protection regulations, the development of ecological tourism, sport and recreation according to the functions of the national park;
7. Prevent degradation of the national park using control and supervision measures, and protect against natural disasters, and seek to enhance the quality of life and the availability of work for the local people.

According to data from the protection and development programme of Tara National Park from the period 2002-2006, it is clear that several challenges affect the management of the National Park. The Park is endangered by the utilization of raw mineral materials, mainly stone and other resources, and also by the exploitation of space for building and tourist purposes without appropriate prior planning or adherence to regulations and construction norms. Other problems include insufficient financial support as well as the lack of support for the creation of a programme for the development of the national park [2]. All the income generated by the public enterprise responsible for the management of Tara is derived from timber. This leads to the question: can and should the income from timber, under a system of restricted fellings (this was, to a high degree, the original intention of the national park designation), finance the realization of almost all of the activities of the national park? [17].

The development concept of the region is based on the utilization of natural resources, with a focus on the preservation of biodiversity and the necessity for tourism and recreation; the production of traditional and healthy food; the establishment of small handicrafts, especially in the protected zones; protection of natural resources and biodiversity involving the application of necessary sanitation and reconstruction measures, and the engagement of labour from the surrounding villages in the activities of the national park. It is necessary to get institutional support for all of these development activities in Tara National Park [2].

Based on research conducted in 2004 on Tara National Park, the findings indicate the need to strengthen the clarity of nature conservation policy and the missions of the responsible authorities. In addition, in order to promote the involvement of local people and empower the national park management to work with them collaboratively, it is necessary to promote communication among all stakeholders [3, 18-20].

2. Methodology

During the research, two types of data were collected: primary and secondary data. Primary data were obtained through the survey method where a questionnaire was devised to enable direct communication with the respondents.

The purpose of the questionnaire involved the determination of economically unusable values of the park by the travel-cost method, determination of the natural values of the park and understanding the relationship between the tourists and the protected area. Also, the aim was to detect the elements that influence the formation of a positive or negative attitude towards the natural values of the park and conservation of nature and species.

The sample consisted of 60 visitors, who stayed in the Tara National Park, in the tourist locality Kaludjerske Bare, in January 2009. On average, each interview lasted about 15 minutes involving individuals, groups and families. The survey questionnaire included a mixture of open, fixed-response and multiple response questions. A combination of mixture was used to examine the various dimensions to the respondents' attitudes and especially to get right information [13,21].

The first set of questions was related to socio-demographic characteristics: gender, age, number of members in the household, education, occupation and monthly income per household. In the second set of questions, the respondents answered which type of transport they had used to reach Tara, how long they planned to stay, how many times they had visited the park, where they came from and how much money they had spent on the trip to Tara and during their stay on the mountain. Questions were also asked relating to whether the respondents were aware of the fact that Mt. Tara is a national park and whether they knew why Tara had been designated a national park. Also, the respondents were asked to answer which capacities of the park they had visited and which are, in their opinion, activities of priority. Within this set there are questions about the level of awareness, opinions, motivations and aspirations. This part of the survey is also the most comprehensive since it is a study of attitudes and opinions. For the better understanding of the relationships between tourists and nature, the tourists gave their answers regarding the questions which consisted of a series of statements, where they expressed the degree of agreeing or disagreeing. The statements were related to how important it is to preserve the environment, to what extent they are willing to give up their personal pleasures in favor of endangered species in the Tara NP, and whether they are willing to support the preservation and protection of species and nature through their willingness to allocate certain funds for this purpose.

In the analysis, we used qualitative and quantitative approach. Quantitative research attempts to explain social reality by means of controlled, mathematical methods [22]. The basic goals include the quantifying of social phenomena, the formulating and testing of theories, and the making of predictions [23,24]. Qualitative research seeks to understand social phenomena, where the researcher is interested in the values and feelings that determine human actions in certain situations.

The shortcomings of the research are related to the short time period in which the research was done. Also, the research was done only in one locality and the size of the sample was inadequate due to the limited budget. The research should be repeated over a longer period of time, continuously during the summer months and at other tourist sites in the area of Tara and thus obtain more comprehensive data on the behavior of tourists in the Tara NP and their attitude towards nature.

Secondary data includes relevant documentation, such as written reports and programmes provided by the public enterprises, a spatial plan of Tara NP, reports by the Institute for Nature Protection of Serbia and reports by the Ministry of Environment, Mining and Spatial Planning.

3. Results

3.1. Demographic and socio-economic structure of tourists

The research included 60 visitors of NP Tara and was done at the site called "Kaludjerske Bare". There were 38 male and 22 female respondents (Table 1). Table 1. shows the age structure of the respondents.

In regard to education, the majority of the respondents are highly educated (Table 1.). Regarding the monthly income at the household level, most respondents did not want to declare themselves (24) and the obtained answers are shown in Table 1.

Gender	Frequency	Percentage
Male	38	63.3
Female	22	36.7
Total number of interviewed tourists	60	100.0
Age structure of the respondents	Frequency	Percentage
Less than 24	5	8.3
25-34	28	46.7
35-44	12	20.0
45-54	10	16.7
55-64	5	8.3
More than 64	/	/
Total number of interviewed tourists	60	100.0
Education	Frequency	Percentage
Primary school	2	3.3
High school	15	25.0
Two or three-year college	10	16.7
Faculty	28	46.7
Postgraduate studies	5	8.3
Total number of interviewed tourists	60	100.0
Monthly income per household	Frequency	Percentage
From 15,000 to 30,000 RSD	2	3.3
From 30,001 to 50,000 RSD	2	3.3
From 50,001 to 80,000 RSD	11	18.4
From 80,001 to 100,000 RSD	9	15.0
More than 100,001 RSD	12	20.0
I would rather not say	24	40.0
Total number of tourists surveyed	60	100.0

Table 1. Research results concerning demographic and socio-economic structure of the tourists interviewed

3.2. Tourists familiarity with the NP Tara

Health and leisure features of tourism travel are incorporated in the tourists' values to which tourist demand is aimed at, especially because the tourists' desire to travel is often identified with a desire for rest, entertainment and recreation in an ecologically healthy environment [25]. Tourism enables man to perceive the preserved richness of nature on the spot. This way of getting to know the broad layers of the society is a kind of "school in nature" and should be nurtured and expanded [26]. The importance of Mountain Tara for the development of tourism has increased in particular after the mountain was declared a national park and after the construction of accommodation facilities [1].

The research results indicate that (96.7%) tourists are aware of the fact that Mountain Tara has the status of National Park, only (3.3%) tourists gave a negative answer.

Further on, when tourists were asked to answer why Mountain Tara was declared a national park, the obtained answers point out the fact that it is because of: 'untouched nature', 'beautiful scenery', 'preserved nature', 'rare and endangered plant and animal species', many of which mentioned Pančić's Spruce, 'rich wildlife', 'lakes' and 'forest' (Figure 5).

Figure 5. Landscape in Tara NP

3.3. Travel expenses

In the last forty years, methods have been developed for assessing the economic value of non-market natural resources for parks and recreational areas. The method of travel expenses is often used to assess the recreational value of protected areas [27]. The method is based on the evaluation of total costs incurred by a visitor within the protected area that reflect in the willingness of a visitor to accept the costs of maintaining the protected resources by paying for them. One should also know that the results of applying this method are typical for each region [28].

Travel expenses represent the sum of transport costs and the expenses of staying at the desired destination. The assumption is that age, educational level and monthly household income affect travel expenses. The expected impact is directly proportional to the increase in costs.

The correlation between the distance from the place of residence and Mountain Tara corresponds to the costs of transport. Those tourists who come from Vojvodina or Southern Serbia spent more money on transport to Tara than those living closer to Tara. Also, tourists who came by car spent more money than those who came by bus.

The length of stay on the mountain is parallel with the increase in total costs. In accordance with the increase of the number of days spent on the mountain, the amount of money spent by a tourist also increases, and the length of stay itself corresponds to the entire experience and impression that a tourist gets about NP Tara.

The summarized results show that almost a third of the tourists interviewed was for the first time on Tara (30%) while, on the other hand, about a third of the respondents (26.7%), visited NP Tara for more than 4 times. The remaining number of the respondents visited Tara from 2 to 4 times. Among other things, the number of visits to the park has an influence on the creation of positive attitudes towards nature conservation; with an increasing number of visits the intensity of experience in nature also increases which then strengthens the relationship man-nature, or man- the environment and vice versa (Figure 6.).

Figure 6. Tourists in Tara NP

3.4 The correlation between visitors and the natural and tourist values of NP Tara

Many national parks were originally declared as places of inspiration for the public, and for their spiritual and physical recovery [29]. Thus, tourism has become a powerful tool in the function of human health – physical, mental and psychological [30,31].

Research results presented in Table 2. refer to the activities in the park that tourists undertook during their stay on Tara.

	Frequency
a) visit to the central part of the national park	17
b) walking tours through nature, through attractive regions	42
c) children's programs and activities	8
d) sports events	6
e) educational and cultural events	13
f) other	2
Total number of interviewed tourists	60

Table 2. Research results concerning the activities in the park that tourists undertook during their stay

3.5. Relation of tourist towards the protection and preservation of nature in Tara NP

Protected areas are essential for maintaining healthy environment. Those areas provide higher quality of life and opportunity for recreational activities [32].

The preserved nature is a major resource for tourism development. The natural environment is necessary to preserve, protect and organize in order to maintain the natural and tourism values. However, with the development of tourism inevitably occur degrading processes of nature. The extent to which are tourists aware of their negative influence on nature, what is their attitude towards nature protection and protection of species, and how willing are they to support protection, will be elaborated in the following analysis.

Statement that relates to the amount of fees in NP Tara "I have already paid too much taxes" represent extreme reluctance of tourists, as 40.0% of respondents evaluated this statement under number 3. After this "neutral category", 35.0% of respondents belong to category "reject completely". Similar category is "somewhat reject", under the number 4, which was supported by 6.7% of respondents. Category "agree fully" support 11.7% respondents, while "somewhat agree" category choose 6.7% of respondents.

Figure 7. illustrates the level of agreement with statement "Protection and conservation of nature is a very important issue and I think that the costs of those services are above the other costs according to their importance".

National parks have many functions, and touristic and recreational are one of the most important between those. In order to preserve and maintain the fundamental phenomenon

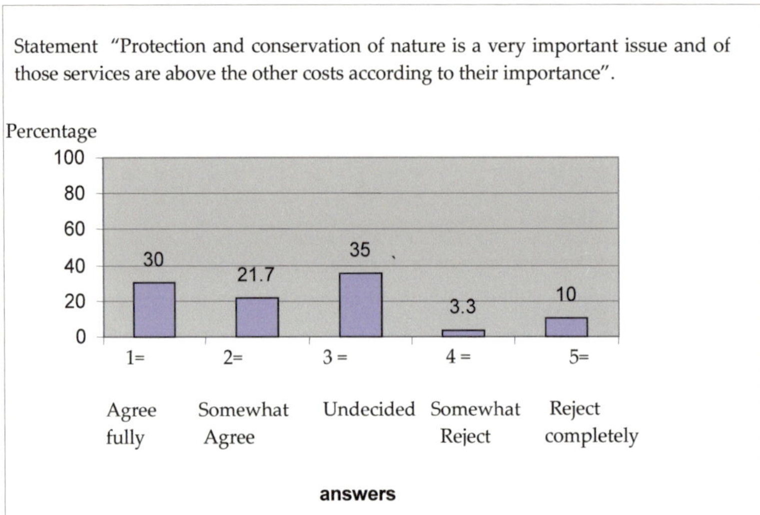

Figure 7. The survey results pertaining to direction and intensity attitude which tourists have according to nature conservation

because of which territory got the status of national park, comprehensive understanding of the nature is necessary (Figure 8.) [26].

Figure 8. Beautiful scenery in Tara NP

The extent to which tourists are aware of their, very often, degrading effect on the protected area, as well the extent to which they are ready to give up personal pleasures in favor of protecting endangered species is shown in Table 3. Construction of infrastructure and material base for tourism leads to serious damage of vital elements of the environment, the loss of natural resources, and therefore the impairment of recreational attributes and attractive environment.

Statement	Assessment scale	frequency	percentage
Among others (e.g. conservation nature and species), a major task of the National Park is the provision of the visitor infrastructure. Increasing the number of visitors due to these activities might threaten the existence of endangered species. To which extent would you enjoy extended visitors' infrastructure even if it would threaten conservation endangered species? (value with 1=I would use facilities extensively, to 5=I would not use facilities at all))	1= I would use tourist facilities extensively	9	15.0
	2= I would use tourist facilities to certain extent	1	1.6
	3= undecided	22	36.7
	4= I would not use tourist facilities to certain extent	10	16.7
	5= I would not use tourist facilities at all	18	30.0
Total number of interviewed tourists		60	100.0

Table 3. The survey results pertaining to extent of willingness tourists to discard tourists' applicability for conservation endangered species

The most of respondents (36.7%) expressed a neutral position regarding the willingness to forego for the sake of expanding tourism in contrast to survival of endangered species. Position under the number 5 defines a waiver of touristic facilities in order to preserve endagered species, for which opted 30.0% of respondents. Close to previous oppinion is the category under the number 4, which point out that tourists would use touristic facilities to some extent, for which 16.7% of respondets gave a vote. The percentage of respondents who would use the maximum of tourist facilities, regardless of the consequences on nature, is 15.0%, following with 1.6% of respondents who would used a those amenities to some extent.

4. Discussion

National parks have the privilege to be the highest in the ranking of protected natural resources, because of their most valuable and most attractive natural resources and quality.

However, it is wrong to think that national park just fulfills the function of relaxation and recreation as other touristic places. Primarily, the set of rules regarding proper behavior and lifestyle should exist in the national park. Protected area should be used for sightseeing and enjoying in nature, and tourist visits must be subordinated to the principles of nature protection [26].

Tourists who were the part of the survey in framework of this research are aware of the fact that they are located in highly valuable natural environment, but their understanding of the essence of nature protection is not developed enough. Tourists have extensively cited the features which they consider as reasons for the designation of NP Tara. Although those characteristics are quite general and can refer to any protected area, surveyed tourists still showed the certain knowledge about the natural environment, which is a good base for further education and encouraging programs for the raising awareness about environment.

The results indicate indefinite attitude of tourists towards the issue of nature protection in Tara National Park, which were showed by opting mostly for neutral statements and by not expressing a clear opinion. It is evident that spatial attention should be given to this group of tourists through educational programs and facilities that would expand their knowledge about the relationship man-nature. Education would contribute to better relationship of a wider range of tourists with NP Tara and developing more humane and ecological relationship of visitors with nature. It is encouraging fact that the part of tourists who declared positively to the protection of nature represents the next major category of respondents. They have better sense of the need for nature protection, as well as sense for aesthetic of nature and its beauty. Of course, it must be emphasized that there is also the part of tourists who were negatively declared toward the protection of nature and its values. It is a good indication that this category of visitors is quite small, and therefore it is necessary to facilitate new ways of thinking and raising awareness of the value and importance of Tara National Park.

Research showed that tourists have mostly same opinion regarding the evaluation of touristic offer, where the high values are given to recreation in nature, long walks, diverse sport activities, but also for sightseeing of animals and plants. It is clear that these results are a kind of advice for management of NP Tara. It indicates the need for diversifying of offers, which would respond to the demands of visitors and in the same time be in line with nature protection. A thoughtfully organized touristic offer with specific facilities and ecological image would present the Tara National Park in the best light, as an institution of protection and tourism.

5. Conclusions

Protected areas are primarily viewed in biological or ecological terms, but some scholars emphasised the economic importance of land managed for conservation objectives. However, protected areas are important at many levels, from local and national to global levels, and they carry out numerous functions beneficial to humans, and even essential to human welfare [33]. Protected areas contribute to a country's social and economic objectives through supporting ecosystem services, promoting the sustainable use of renewable resources, as well as providing places for tourism and recreation. Tourism is considered to be a viable conservation strategy which can benefit the local communities and contribute to the conservation of the protected areas.

Advantages and motives of people staying in protected areas are: enjoying in the simplicity of life in a protected area, awareness of the beauty of protected area; feeling better and making a connection to the natural environment; sharing of these experiences with other people, getting to know the plants and animals that extinct from urban areas; activating the muscles after long walks, spending the time in an environment that allows the deepening of life issues and dilemmas [33].

Tourists visit the national parks and other protected areas, because in them they are able to experience the values that cannot be found elsewhere. The role that national parks play in society and in tourism is often associated with national identity [29]. National Parks provide the desired physical and mental peace, relaxation and aesthetic experience, which inspire a man to "return to nature" [34].

Natural rarities of Tara National Park with many attractive features are of paramount importance for tourism, which represent an opportunity for development of various forms of tourism that should aim to preserve the ecological determinant in the area. The economic benefits from tourism impose the necessity for establishing a better relationship between the visitors and the protected area. Tourism is one of the most favorable economic activities in NP Tara, and the protection of environmental resources is the only guarantee for the realization of tourist activities.

The paper contributes to a better understanding of the relationship of tourists and the recreation values of park, nature protection and the determination of natural values of the park. The results indicate that socio-demographic factors and the number of visits to the park affect the formation of attitudes toward visitors of NP Tara. Those tourists who have more income, higher level of education and who have repeatedly visited NP Tara demonstrated a positive attitude towards nature conservation in the park. The majority of respondents did not express the attitude toward the nature protection (over the half of respondents) and they are characterized by belonging to various socio-economic categories.

Suggestion for the park managers and government agencies is to focus on development of programs to raise environmental awareness of tourists in order to better understand the protection and preservation of the Tara National Park. For the sake of more efficient management strategies it is needed to support activities prioritized by tourists, and that are in accordance with the preservation and protection of the environment.

Introducing the selling the entrance ticket for the park is quite reasonable and practical, which was confirmed with the result of survey. Therefore tourism promotes the nature protection and has an active function, while a protection gain the necessary financial resources for its functioning.

This survey could serve to the administration of the National Park Tara in considering the options to improve the awareness of tourists in terms of better understanding the natural values of protected areas, and with the purpose of improvement of management of natural resources of this protected areas.

Author details

Jelena Tomićević, Ivana Bjedov, Ivana Gudurić and Dragica Obratov-Petković
Department of Landscape Architecture and Horticulture, University of Belgrade - Faculty of Forestry, Belgrade, Serbia

Margaret A. Shannon
FOPER II - European Forest Institute, Varaždin, Croatia

6. References

[1] Mosurović M., Simić M. Tara-Turistički vodič. Užice: Grafos; 2002.

[2] Javno preduzeće 'Nacionalni park Tara' Program zaštite i razvoja područja Nacionalnog parka Tara za period 2002-2006 godina. Bajina Bašta: Javno preduzeće 'Nacionalni park Tara'; 2002a

[3] Tomićević J., Shannon M.A., Milovanović M. Socio-economic impacts on the attitudes towards conservation of natural resources: Case study from Serbia. Forest Policy and Economics 2010; 12: 157-162.

[4] Gajić M. Flora Nacionalnog Parka Tara. Beograd: Šumarski fakultet i Šumarska sekcija Bajina Bašta; 1989.

[5] Tomićević, J., Milovanović, M. Park - people relationship: A case study from Serbia. Proceedings of the International Conference -1st International conference 'Research people and actual tasks on multidisciplinary sciences', 6-8 June 2007, Volume 3, ", Lozenec, Bulgaria, Bulgarian National Society of Agricultural Engineers Engineering and Research for Agriculture; 2007. p91-96.

[6] Tomićević J., Milovanović M., Konold W. Uloga participacije lokalne zajednice u održivom korišćenju prirodnih resursa Nacionalnog parka Tara. Udruženje šumarskih iženjera i tehničara Srbije, Beograd, Šumarstvo 2005; (4) 81-92.

[7] Zavod za zaštitu prirode Srbije Stanje područja nacionalnog parka Tara. Beograd: Zavod za zaštitu prirode Srbije; 2002.

[8] Javno preduzeće 'Nacionalni park Tara' Srednjoročni plan izrade i donošenja programa gazdovanja šumama sa pravom svojine na teritoriji opštine Bajina Bašta. 2002 – 2007. godina. Bajina Bašta: Javno preduzeće 'Nacionalni park Tara'; 2002b.

[9] Stevanović V. The red data book of flora of Serbia 1, Extinct and Critically Endangered Taxa. Ministry of Environment of the Republic of Serbia, Faculty of Biology, University of Belgrade, Belgrade: Institution for protection of Nature of the Republic of Serbia; 1999.

[10] Gajić M., Kojić M., Karadžić D., Vasiljević M., Stanić M. Vegetacija Nacionalnog Parka Tara. Beograd: Šumarski fakultet; 1992.

[11] Obratov D., Djukić M. Flora Diversity of the National Park Tara. In: Forest Ecosistems on the National Parks. International Scientific Conference held at Tara National Park, Bajina Bašta, Serbia/Yugoslavia, September 9th-12th, Monograph on the Subject Inclusive of The Conference Report, Belgrade: Ministry of Environment of the Republic of Serbia; 1996. p226-229.

[12] Institute for nature conservation Proposal to support the Tara Mountain Biosphere Reserve nomination. Belgrade: Institute for nature conservation; 2003.

[13] Tomićević, J. Towards Participatory Management: Linking People, Resources and Management. A Socio-Economic Study of Tara National Park. Culterra, Schriftenreihe des Instituts für Landespflege der Albert-Ludwigs-Universität Freiburg, Heft 43, Freiburg; 2005.

[14] Tomićević J. The impact of benefits on conservation attitudes of local people in the Tara National Park. „Management of forest ecosystems in national parks and other protected areas". Proceedings, Jahorina – NP Sutjeska, July 5-8, Jahorina: 2006. p485-492.

[15] Nikolić S. Turizam u zaštićenim prirodnim dobrima Srbije. Beograd: Zavod za zaštitu prirode Srbije; 2006.

[16] UNCE Prikaz Stanja Životne Sredine u Srbiji 2002 godine (prevod). Beograd: Ministarstvo za zaštitu prirodnih bogastava i životne sredine, Republika Srbija; 2003.

[17] Vučković M., Ranković N., Stamenković V. Relation of Ecological and Economic Function of the Forest of the National Park "Tara". In: Forest Ecosistems on the National Parks. International Scientific Conference held at Tara National Park, Bajina Bašta, Serbia/Yugoslavia, September 9th-12th, Monograph on the Subject Inclusive of The Conference Report, Belgrade: Ministry of Environment of the Republic of Serbia; 1996. p169-174.

[18] Tomićević J., Shannon M. A., Vuletić D. Developing Local Capacity for Participatory Management of Protected Areas: The Case of Tara National Park. Šumarski list, 2010; 134 (9-10) 503-515.

[19] Tomićević J., Shannon M.A., Bjedov I., Obratov-Petković D. Linking protected areas management and local communities – example of Serbia, First Serbian Forestry Congress (under slogan - Future with Forests) -On the occasion of jubilee marking 90 years of its educational, scientific and professional work, Forestry faculty organises the international scientific conference, In: Ristić, R., Medarević, M., Popović, Z. (eds.) (2010): Congress Proceedings, November 11-13, Belgrade: Faculty of Forestry, University of Belgrade; 2010. p1124-1132.

[20] Tomićević J., Bjedov I., Shannon M.A., Obratov-Petković D.Understanding Linkages Between Public Participation and Management of Protected Areas-Case Study of Serbia. In: Ishwaran N. (ed.) The Biosphere. Rijeka: InTech; 2012. p131-142.

[21] Tomićević J., Bjedov I., Obratov-Petković D., Milovanović M. Exploring the park-people relation: collection of Vaccinium myrtillus l. by local people from Kopaonik National park in Serbia. Environmental management 2011; 48 (4) 835-846.

[22] Atteslander P. Methoden der empirishen Sozialforschung. 8. Auflage Sammlung Göschen. Berlin: de Gruyter; 1995.

[23] Ragin C. Constructing Social Research, Pine Forge Press: Thousand Oaks; 1994.

[24] Flick U. Qualitative Forschung: Theorie, Methoden, Anwendung in Psychologie und Socialwissenshaften, 2. Auflage. Hamburg: Rowohlts enzyclopädie; 1996.

[25] Dojčinović – Đukić V. Kulturni turizam. Beograd: Klio; 2005.

[26] Vidaković P. Nacionalni parkovi i turizam, Zavod za zaštitu SR Hrvatske, Zagreb: Insitut za turizam u Zagrebu; 1998.

[27] Ortaçeşme V., Özkan B., Karagüzel O. An Estimation of the Recreational Use of Kursunlu Waterfall Nature Park by the Travel Cost Method, Tubitak, Turkey, Turk J Agric For 2002; (26) 57-62.

[28] Dixon J.A., Pagiola S. Economic Analysis and Environmental Assessment. Environmental Economics and Indicators Units, Environment Divelopment; 1998.

[29] Eagles P.F.J, McCool S.F. Tourism in National Parks and Protected Areas and Managment. UK: CABI Publishing; 2002.

[30] Savelli A. Sociologia del turismo. Milano: Franco Angeli; 1993.

[31] Tomićević J., Milovanović M. Status zaštićenih područja i ciljevi upravljanja. Udruženje šumarskih iženjera i tehničara Srbije, Beograd, Šumarstvo 2006; (1-2) 181-188.

[32] CBD Protected Areas in Today's World: Their Values and Benefits for the Welfare of the Planet, CBD Tehnical Series No. 36, Montreal: Secretariat of the CBD; 2008.

[33] Harman D., Putney A.D. The Full Value of Parks. Oxford, UK: Rowman & Littlefield Publisher,INC; 2003.

[34] Nikolić S. Priroda i turizam Srbije. Beograd: Zavod za zaštitu prirode Srbije i Eko-centar; 1998.

Chapter 5

Development Prospects of the Protected Areas System in Croatia

Ivan Martinić, Barbara Sladonja and Elvis Zahtila

Additional information is available at the end of the chapter

http://dx.doi.org/10.5772/50482

1. Introduction

Creation of protected areas is the most common approach to conserve global biodiversity (Fu et al., 2004). Protected areas are places with extraordinary biological value providing numerous functions to humans (Tomićević et al., 2011). In recent years, management of protected areas has become one of the relevant aspects in national and international studies dealing with nature conservation and management (Muñoz-Santos and Benayas, 2012).

Protected areas in a contemporary social context have an important role and responsibility in Croatia. Croatia is attempting a popular participatory approach to nature protection (Kapoor, 2001; Khadka & Nepal, 2010; Parker & Thapa, 2011; Robertson and Lawes, 2005; Sladonja et al., 2012). Protected areas' role is viewed through the fulfillment of the objectives of biodiversity preservation, but also through a full contribution to sustainable development and especially the economies of local communities. The most important features of the approach to managing protected areas in Croatia, together with strategic directions for achieving the goal of further protected areas system development and the increase of their effective management with the active participation of the public must be identified. The key problems of daily functioning of protected areas in Croatia refer to undifferentiated and lacking funding, inconsistency of regulations and legal ambiguities, the lack of a central National Park Agency on "one vision, one mission" approach to managing and the low level of implementation of so far adopted management plans of protected areas.

Prospect of park system development in Croatia is perceived in three key aspects: 1) improving the functioning, 2) contribution to sustainable development and 3) sufficient long-term financing. Achievement generator of these perspectives should be the future National Park Agency for Croatian protected areas (NPA) with the authority to define and implement a unitary park policy and privileges of presenting overall interests and needs of the protected areas system according to the political and professional environment.

The perspective of contribution to sustainable development is related to the establishment of an organized system of visiting the protected areas at national level and implementation of multi-day program of visits – both with significant involvement of local communities. Addition of the tourism development strategy in protected areas is welcome, with emphasis on the unique marketing and widespread use of modern technology in all aspects. The perspective of sustainable funding is based on the current unfavorable income and expenditure of protected areas where almost 85% of revenue is channeled to the costs of employees and current operations, and only 15% on the investments and program costs, which directly affects the volume of activities for reinforcement of local community economies. The calculation of the financial sustainability of protected areas is appreciated and included through the direct and indirect market values, but also non-market and non-usage values and benefits of protected areas as a condition of knowing the total value and actual costs of protection, preservation and development.

1.1. Overview of nature protection legislative in Croatia

The first examples of nature protection in Croatia are linked to the second half of the 19th century. The first institutions that were funded with the aim of nature protection were the Croatian Nature Society (1885) and the Society for the Arrangement of Plitvice Lakes and its Surroundings (1893) (Sladonja et al. 2012). The first official framework for nature protection was set by issuing of the Law for Bird Protection (1893), Law on Hunting (1893) and Law on Underground Protection (1900), while a comprehensive legal framework in the form of the Law on Nature Protection was finally completed in 1960 during the period of communism in the former Socialist Federal Republic of Yugoslavia (SFRJ). Croatia has existed as an independent State since 1991. In the last two decades, nature protection service often changed positions in governmental institutions. From 1990-1994 it was under the Ministry of Environmental Protection, Physical Planning and Construction. From 1995, it was under the State Directorate for the Protection of Cultural and Natural Heritage and then from 1997 under the Ministry of Culture. In this same year the service entered into the system of the State Directorate for Nature and Environmental Protection and from 2000 it was part of the Ministry of Environmental Protection and Physical Planning. The Nature Protection Directorate has been an integral part of the Ministry of Culture since 2004. By coming into force of the Act on the Organization and Scope of Work of Ministries and Other Central State Administration Bodies (Official Gazzette (OG) No. 150/11 and 22/12), adopted by the Croatian Parliament at its session held on 22 December 2011, the management of nature conservation has been taken over by the Ministry of Environmental and Nature Protection. In 2002, the State Institute for Nature Protection (SINP) was established by the Government Regulation as the central expert institution for nature protection. It is the central institute dealing with expert tasks of nature conservation in Croatia. The Institute was established according to National Strategy and Action Plan (OG 81/99) and Implementation Plan on Stabilization and Association Agreement. The State Institute for Nature Protection carried out a series of activities aimed at ensuring the lasting conservation and improvement of Croatia's natural heritage (State Institute of Nature Protection, 2012).

1.2. Status and management of protected areas in Croatia

Protected areas in Croatia account for 8.54% of the total area of the Republic of Croatia, or 11,38% of its land area. From this 4,76% are National parks and Nature parks. Up to now in the Register of protected natural assets of the Republic of Croatia 461 protected areas are registered, 9 of which preventatively protected (Table 1). According to The Nature Protection Act (OG 70/05 and 139/08) in Croatia there are 9 national categories of protection, aligned with IUCN categories (Table 2).

Category	Number	Surface area (km²)
Strict reserves	2	23,95
National park	8	961,35
Special reserve	83 (4)*	853,34
Nature park	11	4.242,15
Regional park	2 (2)	1.599,91
Nature monument	116	2,46
Significant landscape	79 (1)	880,75
Park forest	36 (1)	88,89
Monuments of park architecture	122 (1)	9,56
Total	**459 (1)**	**8.662,46**
Protected areas inside other protected areas		1.205,15
Total surface area of protected areas in Croatia		**7.457,31**

*in brackets are preventatively protected areas

Source: Strategy and action plan for biological and landscape diversity protection of the Republic of Croatia, Ministry of Culture, Republic of Croatia, 2010

Table 1. Review of protected areas in Croatia – number and surface areas according to categories

Protected areas in Croatia are managed by the public institutions for the management of protected natural areas. The basic goal of their activity is the management of protected areas, in the sense of protection, maintenance and promotion, ensuring the unhindered unfolding of natural processes, and sustainable use of natural resources. Public institutions of National and Nature parks are established by virtue of a Regulation of the Government of the Republic of Croatia. Public institutes for the management of other protected areas are established by local or regional self-government units. Decentralization on nature protection started in the 2005 with issue of the new Law on Nature Protection (OG 70/05). Today present regional and local institutions are a direct result of this law (Sladonja et al., 2012). Counties may hand over the management of a protected area to the local self-government unit, i.e. to a public institute established by a town or municipality. In Croatia, there are currently 19 public institutes at the national level, 20 at the county level and 7 at the local level (State Institute for Nature Protection, 2012).

Protection category	Intent	IUCN category	Manag. level
Strict reserve	Conserve intact nature, monitor the state of nature and education	I	county
National park	Conserve intact natural values, scientific, cultural, education and recreation intent	II	national
Special reserve	Conservation due to its uniqueness, rarity or representativeness, and of particular scientific significance	I/IV	county
Nature park	Protection of biological and landscape diversity, education, cultural, historical, tourism, recreation intent	V/VI	national
Regional park	Protection of landscape diversity, sustainable development and tourism	V/VI	county
Nature monument	Ecological, scientific, aesthetic or educational intent	III	county
Significant landscape	Conservation of landscape values and biological diversity, or cultural and historical values or landscape of preserve unique characteristics, and for rest and recreation	V	county
Park-forest	Conservation of natural or planted forests of greater landscape value, rest and recreation	V	county
Park architecture monument	Conservation of artificially developed areas or trees having aesthetic, stylistic, artistic, cultural, historic, ecological or scientific values	No adequate IUSN category	county

Table 2. The national categories of protected areas according to The Nature Protection Act (OG 70/05 and 139/08)

National parks in Croatia are defined by the Law on Nature Protection, (OG 70/05, Article 11), as large land and/or aquatic areas mostly non modified, with exceptional and multiple nature values, embracing one or more preserved or slightly modified ecosystems, primarily aimed to natural genuine values preserving. From 8 National Parks in Croatia, 3 are marine areas (Brijuni, Kornati and Mljet), 3 mountain areas (Risnjak, North Velebit and Paklenica) and 2 freshwater sites (Krka and Plitvice Lakes). Plitvice Lakes National Park, designated in 1949. is the oldest park in Croatia and today is the most visited park, while the most recent one is North Velebit (Fig. 1.).

Nature parks are natural or partially cultivated land or aquatic areas with ecological features of national or international significance, with accentuated landscape, educational, cultural, historic, touristic and recreational values. Today there are 11 Nature parks in Croatia spread all over the territory (Fig. 1.).

Protected areas in Croatia, due to their special values make the core and bases for biodiversity and landscape protection, and are key points of national ecological network as well as of future ecological network NATURA 2000 in Croatia. NATURA 2000 is the EU ecological network composed of the most significant areas for conservation of species and habitat types. After the accession of the Republic of Croatia into the European Union, NATURA 2000 will also be proclaimed in the territory of our country. In the meantime a preparation project EU Natura 2000 Integration Project (NIP) is going on. Based on the Loan Agreement with the World Bank (IBRD 8021-HR) signed on 22nd February 2011, ratified by the Law on Ratification of the Loan Agreement (OG MU 7/2011 from May 18, 2011) Ministry of Culture began implementing a five-year EU Natura 2000 Integration Project (NIP). Due to new Act on the Organization and Scope of Work of Ministries and Other Central State Administration Bodies (OG No. 150/11), from December 22, 2011, Nature Protection Directorate and NIP accordingly, are under jurisdiction of new Ministry of Environmental and Nature Protection. The objectives of NIP are to:

- help in supporting National parks, Nature parks and County Public Institutions for management of protected natural values to implement European ecological network Natura 2000 objectives in investment programs;
- strengthen capacity for biodiversity monitoring and EU-compliant reporting; and
- introduce programs that involve a wide group of stakeholders in Natura 2000 network management.

In the 2009, WWF initiated the Report on the Representativeness of Protected Areas in Dinaric Ecoregion (Albania, Bosnia and Herzegovina, Croatia, Slovenia and Montenegro) within the Project Protected Areas for Alive Planet – Dinaric Ecoregion (WWF Project, 2010). This project gives a wider prospective on regional biodiversity which was so far performed only on the national basis. Within this Project, analysis of biodiversity aims in Croatia points to some shortfalls. Target species and habitats non adequately represented in protected areas are considered as "blanks". According to Project results Croatia has "blanks" in protection of plain and hilly areas (altitude 0-800 m), carstic fields and reptiles and freshwater fishes. However, it was established that Croatia has the highest 58,6% of aims adequately included inside protected areas compared to the region average of 34,2%. The whole Dinaric region demands for joined activities since the threshold set by IUCN of 10% of protected areas is reached only for land areas in Croatia. Marine protected areas are far below this margin in the whole area.

On the other hand, considering its number and diversity, protected areas in Croatia have a very important role in tourist concepts shaping. Total economic effects for national and local economies increase on a higher rate than most of other economy branches (Martinić et al., 2009).

Some of the most important approach features for protected areas management in Croatia are:

- legislative framework, conciliated with international standards and praxis, for organization of key issues of protected area management;
- state commitment for administrative and professional support in planning, establishing and management of protected areas through access of crucial professional documents (habitat map and other databases about species and habitats, Red book of endangered species etc);
- imperative directing and controlling role of competent state administrative and professional bodies (Ministry of Culture – Department for Nature Protection, Department for Nature Protection Inspection, State Institute for Nature Protection, other state institutions etc.);
- state commitment for establishing and financing institutions for protected area management of interest for the Republic of Croatia, especially through ensuring state budget financing and issuing of spatial plans for the most relevant categories of protected areas in Croatian parliament.

According to National Strategy and Action Plan (NSAP) for protection of biological and landscape diversity protection of the Republic of Croatia (Ministry of Culture, 2010) which is the basic strategic document for nature protection in Croatia the following goal was defined for protected areas:

> *To continue the development of protected area system, efficient management of protected areas, increase the areas under protection and instigate active participation of interested public.*

In order to achieve the mentioned goal, according to NSAP six strategic directives have been determined:

- create basic documents for the protected area management;
- digitalize borders and continue with revision of existing protected areas;
- valorize, categorize and legally protect singular areas;
- ensure the involvement of interested public;
- improve the system for the protected area management,
- solve legal-assets relations and increase the share of state land inside protected areas.

Key problems in every day functioning of protected areas in Croatia, especially of those on the national level – National parks and Nature parks are:

- lack of targeted budget means for basic and programme activities financing National parks and Nature parks, as a consequence of the wrong positioning of nature protection in relation to environmental protection and other key economy sectors;
- inconsistention of regulation and legal ambiguities affecting directly the impossibility of key questions solving in parks (assets-legal relations, concessions etc);
- lack of the National Park Agency on the state level, which would represent global interests and needs of protected areas system;
- lack of parks functioning standards, especially in relation to staff systematization, including education and advancements of parks staff;

- non adequate spatial planning in parks, especially in the aspect of regional planning upgrading;
- low realization of so far accepted management plans, especially due to the excessive number of action plans requiring non-existent human and financial potentials;
- lack of diversified development models of parks financing, including public-private partnership and concession system (Martinić, 2004a; 2004b).

2. Park system development perspective

Adopting all mentioned we can summaries the park system development perspective in three key aspects:

- improving the functioning;
- contribution to sustainable development;
- sufficient long-term financing.

2.1. Functioning improvement perspectives

Due to its complex ecologic and social roles, and tasks Croatian park system needs to upraise on a higher functioning level. Framework for such higher level is determined by Croatian protected areas system development and especially the part of ecological network NATURA 2000, with long term management documents aimed at area protection, rehabilitation and revitalization actions implementing as well as development plans responding to local population needs and interests.

So far, the management of National parks was performed through park boards established by the Republic of Croatia as public institutions. Each park has its board and the coordination between all parks is ensured by the operation of an advisory council for National parks.

Recently, there are plans for the establishment of the National Park Agency (NPA) for Croatian protected areas. The NPA will be the generator of these activities. The NPA will take over the management of 8 National parks and 11 Nature parks. It will have competencies of defining and implementing of a unitary park policy in order to ensure parks administration and management functioning conforming the principle „one mission one vision". This would guarantee the equable functioning models in all system units, with previous rationalization and transformation of existing park system. NPA should enter in function as soon as the new Law on Nature Protection will be approved by the Croatian Parliament and issued. NPA will become the central address of the whole park system and will ensure the direct communication between parks and state administration (parliament, ministries) and other state institutions, international bodies and organizations. On the protected areas level, key gains would be more effective solutions of assets-legal relations, systematization of occupational and positional staff, further standardization of administration procedures, sufficient and stable financing etc. Redefining of spatial planning concept is a particularly important role of the NPA, as well as development of protected areas contribution to local population, introduction of differential model for public financing including the public-private partnership and concession system development. According to Law, spatial planning for National parks and Nature parks is performed by Croatian parliament. Finally, NPA should establish a permanent education system, professional specialization and advancement of park staff – creating a better social positioning of park professionals. In the NPA organizational and personnel capacities for capital project activities in all parks would be developed, including project preparing for financing from EU structural funds, investment projects managing and implementing of concession applications for touristic- recreational services in parks.

The future Croatian NPA is shaped on the pattern of the USA National Park Service (NPS). NPS is the U.S. federal agency that manages all National parks, many National monuments, and other conservation and historical properties with various title designations. It was created on August 25, 1916, by Congress through the National Park Service Organic Act, it has today 21.989 employees, 397 unites and annual budget of about $2.900 billions (National Park Service, 2012). There are many similar institutions operating in various countries with the same aim, like Canada and France. Parks Canada, also known as the Parks Canada Agency, is the agency of the Government of Canada mandated to protect and present nationally significant natural and cultural heritage, and foster public understanding, appreciation, and enjoyment in ways that ensure their ecological and commemorative integrity for present and future generations. Parks Canada manages 42 National Parks (including seven National Park Reserves), four National Marine Conservation Areas, one National Landmark, and 167 National Historic Sites of Canada. The agency also administers the Canadian Register of Historic Places, a registry of historic sites in Canada (National Parks Canada, 2012).

The French National parks protecting a total area of 3.710 km^2 are coordinated by the government agency Parcs Nationaux de France. Nine National parks make a unique system of National Parks of France (National Parks of France, 2012). However there are many different experiences. Several countries don't have a unique institution managing National parks and each protected area is maintained by a separate independent institution. There are no firm rules for a guarantee of successful protected area management.

2.2. Sustainable development contribution perspectives

All global social-economic development strategies of the Republic of Croatia consider protected areas to have roles in biodiversity preservation through the contribution to sustainable development. In relation to local communities' progress, it means the possibility to use all protected area qualities with the aim of developing secure, diverse and sustainable workplaces and maintain balanced and healthy local communities.

Although Croatia has a long experience in park policy creating and implementing, there are evident difficulties in achieving roles of protected areas through local communities' contribution (Martinić et al., 2008, 2009). Successful participative conservation requires a legislative framework, management understanding and determination as well as local willingness to implement conservation programs. The traditional top-down approach to nature protection in Croatia is being replaced by bottom-up conservation. This new approach treats local communities as key partners in wildlife management and calls for their participation in social development and biodiversity conservation (Sladonja et al., 2012). So far positive experiences in park policy implementing are linked to the economically most successful National parks such as Plitvice Lakes, Krka and Paklenica, while other protected areas give modest contribution to rural development and local communities' economic welfare. Tourism incomes (entrance fees, shopping, transportation, parking fees, recreation...) can raise environmental awareness and give incentives for a new rationality in traditional resource use (Chen et al., 2005). Countries with a long tradition and successful environmental protection such as Canada, has recently introduced new products and services, like yurt accommodation and innovative interpretive programming. Some local fees may need to be adjusted or established as a result of these service enhancements. All revenues from fees at National parks, National historic sites and National conservation areas are reinvested at the location for such purposes as maintenance, repairs or replace aging visitor facilities as well as enhancing visitor experiences. Prices are set to ensure a good value to visitors as well as a fair economic return for tax payers based on market value and rates (National Parks Canada, 2012).

The most important contribution to local communities is linked to the development of specific tourist functions in protected areas. Considering global trends, the basis for such programs development is the establishment of protected areas organized system of visiting on the national level and implementing of multi-day visiting programs. It is expected that in most protected areas, and not only in those protected on the national level, museums, visitor centers, ecological education centers etc. will become an inevitable part of tourist offer. Multi-day visits would be organized for groups up to 25 participants with professional guidance through thematic programs in 3-5 protected areas with obligatory contact with local population (accommodation, gastronomic offer, craft demonstrations etc). Recent studies have revealed positive perception of local population toward involvement in PAs activities and high support for conservation in Croatia (Sladonja et al. 2012) which is certainly an important element for successful implementation of tourist programs.

Initial examples of multi-day thematic protected area tour in Croatia were developed by Martinić (2002, 2010) through national model called "Nature – Adven(ture)". It is foreseen that visits include visiting parks of "common ecological sign": in the focus of the "green" path is the diversity of forest ecological systems and communities; in the "blue" adven(ture) freshwater habitats are on the path; in the "carstic" carst phenomenon; in the "sea" tour island vegetation and marine ecological systems (Fig. 1.). Marketing would be managed through a shared centre, suggesting a central informative and marketing point of the whole park system where a unique ticket for all expenses on one multi-day tour could be bought.

Realization of these perspectives needs the application of approved Strategy of tourism development in protected areas. It is extremely important that the process of strategy defining and accepting it is done in collaboration with all interested parts in order to satisfy wishes and needs of all stakeholders. Key parts are regional and local authorities, entrepreneurs, NGOs and local population. It's up to them to identify the core of the tourist offer in protected areas and wider.

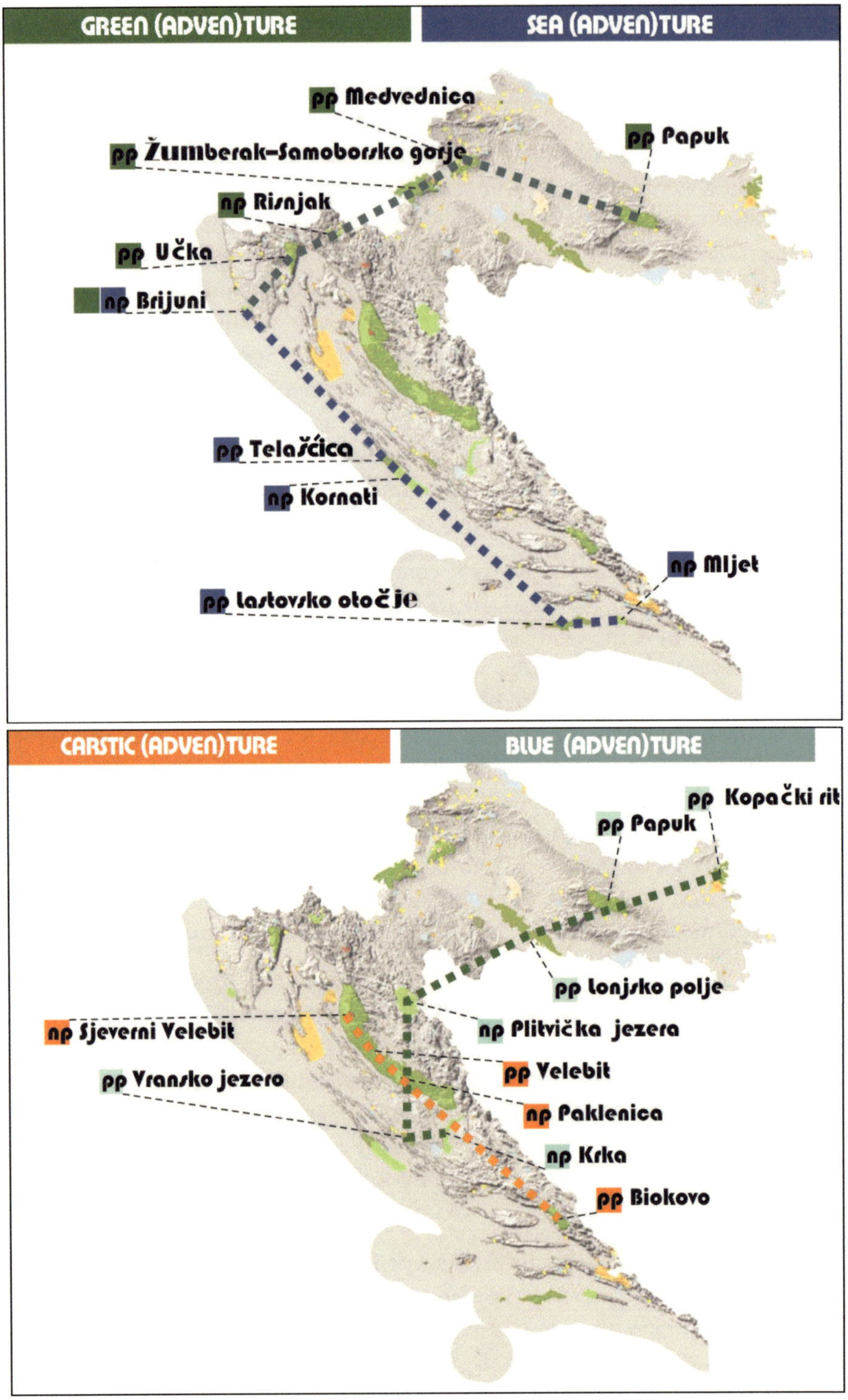

Figure 1. Thematic paths in Croatian protected areas (Martinić, 2010)

2.3. Perspectives of sufficient financing

Sustainability involves balancing ecological, social, and economic development outcomes (Deery et al., 2005; Dwyer 2005; Font and Harris, 2004; Pfueller et al., 2011). The unique character and beauty of protected areas have become attractions for tourism and recreation, it is however important to prevent destruction activities on these sites. Tourism and recreation have a range of damaging impacts on habitats and species (Buckley and Pannell 1990). Even although tourism is a commercial activity requiring economic return to survive, within partnerships with protected area managers, it appears to contribute to sustainability (Macbeth et al., 2004). If tourism is to contribute to sustainable development, then it must be economically viable, ecologically sensitive and culturally appropriate (Wall, 1997). In order to use the tourism in a sustainable way, careful planning and management is required, as well as appropriate budget administration. Budget assignment from national sources to protected areas in Croatia is within global rates. Other incomes of protected areas depend directly from their own possibilities to earn profit from proper activities (entry tickets, tourist services, accommodation, ecological education etc.). In cases of the most known Croatian parks these incomes are very significant and highly exceed the overhead assigned budget.

According to Martinić (2001, 2010) adverse expenses structure in National parks and Nature parks is observed. Even 85% of revenues are channeled to employees' expenses and overheads of the protected areas administration; means that only 15% of incomes are possible to direct in investments and program expenses including activities linked to local communities economy improvement. Low share of program expenses critically diminish possibilities for protected area development and fulfilling their socio-cultural and economic functions. Besides, even preserving the quality of ecological functions is questionable.

In the perspective the calculation of financial sustainability of protected areas is impending by accepting and enclosing not only of their direct and indirect market values but also of their non-market and non-usage values and benefits. Only such structure of economic protected areas will give cognition about its importance, but also real expenses of their protection, preservation and development.

Management concept will have to rely on clear financial mechanisms and concrete financial sources necessary for company functioning and achievements of PAs aims, accepting particularly its commitments linked to the ecological network NATURA 2000.

With this concept, means for working and performing basic activities of park system must be ensured from: state budget for conduction of national protection aims and management of the most valuable nature resources (National parks, Nature parks, Strict reserves) and for performing of international programs for biodiversity preservation (NATURA 2000 etc);

- from other public sources determined by law and special regulations, especially from the Fund from nature protection;
- from charge systems which are usually paid directly or indirectly to park management as entry fees, parking and camping fees;
- from concession fees issued on national level for use of natural resources such water, forest, wild animals, mineral raw materials etc;

- from concession fees according to special contracts with the carriers of tourist-recreational Park activities;
- from tuitions according to special permits for the commercial use of protected areas (promotion, shooting, photographing etc.);
- from souvenirs, maps and books sale;
- from copyright use of protected area sign/logo etc.

Other forms of financial support should be as usually worldwide, more than today based on funds and trusts activities of various agencies, NGOs or structural societies and international projects means. Nowadays, thousands of protected areas in the worlds, especially in developing countries suffer an extreme funding deficit, and many areas have no budget at all (Wilkie et al., 2001). It is necessary to provide rapid actions on global level for stable protected area financing in order to minimize biodiversity loss and promote healthy natural areas as an integral part of sustainable development (Bruner et al., 2004).

3. Conclusions

1. In the modern social context the role and importance of protected areas are observed equally through goals accomplishing in biodiversity preserving and full contribution to sustainable development, and especially to local community economies.
2. Existing park policy concept in the Republic of Croatia gives partial assumptions for further development of protected area system. Their improvement and efficient management with active public participation will contribute to more efficient and sustainable subsistence.
3. Key problems of every-day protection area functioning in Croatia refer to insufficient and non defined financing, legislation and law incapability and partiality, lack of National Park Agency for "one vision, one mission" management approach, weaknesses of spatial planning and low realization of the first generation of management plans.
4. Development perspectives of park system in Croatia can be observed through three key aspects:
   - functioning perspectives,
   - sustainable development contribution,
   - perspectives of sufficient financing.
5. The future National Park Agency for protected areas in Croatia should act as the generator of mentioned perspectives, carrying the responsibility for implementation of a unique park policy and representing global interests and needs of the protected area system.
6. In the close future, real contributions to sustainable development are linked to the establishment of organized protected area visiting system on national level and implementation of overnights visiting programs. The prerogative is issuing of Tourism Development Strategy in protected areas with the accent on unique marketing and wide use of modern technologies.
7. Sustainable financing perspective must be based on calculation of financial sustainability of protected areas considering and embracing direct and indirect market

values and non market and of their non market and non usable values and benefits. Only by this calculation we could acknowledge total values and real costs of protection, preservation and development of protected areas.

Author details

Ivan Martinić
University of Zagreb, Faculty of Forestry, Zagreb, Croatia,

Barbara Sladonja
Institute of Agriculture and Tourism Poreč, Poreč, Croatia

Elvis Zahtila
Natura Histrica, Public Institution for Protected Area Management in Istrian County, Croatia

4. References

Buckley, R. and Pannell, J. (1990). Environmental impacts of tourism and recreation in national parks and conservation reserves. *Journal of Tourism Studies*, 1, pp. 4-32

Bruner, A.G., Gullison, R.E. and Balmford, A. (2004). Financial Costs and Shortfalls of Managing and Expanding Protected-Area Systems in Developing Countries. *BioScience*, Vol. 54 No12, pp. 1119-1126, ISSN 0006-3568

Chen, Z., Yang, J., & Xie, Z. (2005). Economic development of local communities and biodiversity conservation: a case study from Shennongjia National Nature Reserve, China. *Biodiversity and Conservation*, 14, pp. 2095-2108, ISSN 0960-3115

Deery, M., Jago, L. & Fredline, L. (2005). A framework for the development of social and socioeconomic indicators for sustainable tourism in communities. *Tourism Review International*, 9, pp. 66-79, ISSN 1544-2721

Dwyer, L. (2005). Relevance of triple bottom line reporting to achievement of sustainable tourism: a scoping study. *Tourism Review International*, 9, pp. 79-93, ISSN 1544-2721

Font, X. & Harris, C. (2004). Rethinking standards from green to sustainable. *Annals of Tourism Research*, 31, pp. 986-1007

Macbeth, J., Carson, D. & Northcote, J. (2004). Social capital. Tourism and regional development: SPCC as a basis for innovation and sustainability. *Current Issues in Tourism*, 7, pp. 502-522, ISSN 1368-3500

Fu, B., Wang, K., Lu, Y., Liu, S., Ma, K., Chen, L. & Liu, G. (2004). Entangling the Complexity of protected Area Management: The Case of Wolong Biosphere Reserve, Southwestern China. *Environmental Management*, Vol. 33, No. 6, pp. 788-798, ISSN 0364-152X

Kapoor, I. (2001) Towards participatory environmental management. *Journal of Environmental Management*, 63, pp. 269-279, ISSN: 0301-4797

Khadka, D. & Nepal, S.K. (2010). Local Responses to Participatory Conservation in Annapurna Conservation Area, Nepal. *Environmental Management*, 45, pp. 351-362, ISSN 0364-152X

Martinić, I., (2001) Američka iskustva za unapređenje upravljanja nacionalnim parkovima i parkovima prirode u Hrvatskoj, MZOPU RH (izvješće sa studijskog boravka), 1-12, Zagreb

Martinić, I., (2002). Planovi upravljanja za hrvatske nacionalne parkove i parkove prirode. Šumarski list 9-10, CXXVI, 501-509, Zagreb

Martinić, I. (2004a). 55. obljetnica NP Paklenica – Kako osigurati održivost i vitalnost funkcioniranja u svjetlu naglasaka V. svjetskog kongresa nacionalnih parkova. Paklenički zbornik, vol. 2., 147-150. Starigrad-Paklenica, 2004

Martinić, I. (2004b). Ključna pitanja parkovne politike u Hrvatskoj - uz 55. godišnjicu proglašenja prvih hrvatskih nacionalnih parkova, Gazophylacium, Zagreb

Martinić, I., Pletikapić Z., & Kerovec, M. (2008). Realne opcije održivog razvoja u zaštićenim područjima s osvrtom na osmišljavanje održivog razvoja u NP Una. Zbornik radova "Zaštićena područja u funkciji održivog razvoja", 441-460, Bihać

Martinić, I., Pletikapić Z., & Kerovec, M. (2009). Planiranje i upravljanje zaštićenim područjima u funkciji održivog razvoja R. Hrvatske. S stručni skup s međunarodnim sudjelovanjem "Prostorno planiranje, zaštita prirode i okoliša kao pretpostavka gospodarskog razvitka RH na ulasku u EU", Zagreb HGK, objavljeno u "Gospodarstvo i okoliš" 100/09

Martinić, I. (2010). Upravljanje zaštićenim područjima prirode – planiranje, razvoj i održivost. Šumarski fakultet Sveučilišta u Zagrebu

Muñoz-Santos, M. & Benayas, J. (2012). A Proposed Methodology to Assess the Quality of Public Use Management in Protected Areas. *Environmental Management*. Online first DOI 10.1007/s00267-012-9863-0, ISSN 0364-152X

National Park Service, 25.04.2012, Available from: http://en.wikipedia.org/wiki/National_Park_Service

National Parks Canada, 27.05.2012, Available from: http://www.pc.gc.ca/eng/index.aspx

National Parks of France, 27.05.2012, Available from: http://en.wikipedia.org/wiki/National_parks_of_France

Parker, P. & Thapa, B. (2011). Natural Resource Dependency and Decentralized Conservation Within Kanchenjunga Conservation Area Project, Nepal. Environmental Management, Vol. 49, No2, pp. 435-444, ISSN 0364-152X

Pfueller S.L., Lee, D. and Laing, J. (2011). Tourism Partnerships in Protected Areas: Exploring Contributions to Sustainability. *Environmental Management* 48:734-749, ISSN 0364-152X

Robertson, J. & Lawes, M.J. (2005). User perceptions of conservation and participatory management of iGxalingenwa forest, South Africa. *Environmental Conservation*, 32 (1), pp. 64-75, ISSN 0376-8929

Sladonja, B., Bršćić, K., Poljuha, D., Fanuko, N., & Grgurev, M. (2012). Introduction of Participatory Conservation in Croatia, Residents' Perceptions: A Case Study from the Istrian Peninsula. *Environmental Management*. Vol 49. Number 6, June 2012, ISSN 0364-152X

State Institute of Nature Protection, 28.05.2012, Available from: http://www.dzzp.hr/eng/

Tomićević, J., Bjedov, I., Obratov-Petković, D. & Milovanović, M. (2011). Exploring the Park-People Relation: Collection of Vaccinium Myrtillus L. by Local People From Kopaonik National Park in Serbia. *Environmental Management*, 48, no 4, pp. 835-846, ISSN 0364-152X

Wall, G. (1997). Is Ecotourism Sustainable? *Environmental Management*, 21, No4, pp. 483-491, ISSN 0364-152X

WCPA, 2007: Sustainable Tourism in Protected Areas – Guidelines for Planning and Management

Wilkie, D.S., Carpenter, J.F. & Zhang, Q. (2001). The Under-financing of protected areas in the Congo Basin: So many parks and so little willingness to pay. *Biodiversity and Conservation*, 10, pp. 691-709, ISSN 0960-3115

WWF Project Final Report "Protected Areas for Alive Planet – Dinaric Ecoregion" (2010)

Chapter 6

Discrimination of the Decision Structure of Suburban Park Users by Environmental Attitudes

Natalia López-Mosquera, Mercedes Sánchez and Ramo Barrena

Additional information is available at the end of the chapter

http://dx.doi.org/10.5772/50062

1. Introduction

Interest in environmental attitudes arose in the early 1970s. In general, there are several factors such as age, gender, educational status, income and other socio-economic factors for example, see [1] that affect citizens' environmental attitudes. Other factors, which have been included in previous studies, are related to politics or active participation in environmental organizations or the rural/urban character of the place where a household is located, in references [2-3] Working from another standpoint, sociologists and psychologists have developed various theoretical approaches, such as the Theory of Planned Behaviour in [4] and the Value-Belief Norm Theory in [5], to explain how environmental attitudes affect citizens' behaviours. Thus, most research has evaluated the environmental attitudes of individuals towards political participation in different plans of action; environmental conservation and willingness to pay for the use and conservation of different natural landscapes, see [6-11]. Therefore, environmental literature has focused on explaining the factors which condition individuals' environmental attitudes and the effect of environmental attitudes on different Behaviours have been widely studied. However, no studies have analyzed how environmental attitudes determine individuals' decision structure. This study takes 'decision-making' to be the cognitive process leading towards a decision.

Recent studies have focused on factors affecting the decision structure of individuals to various natural landscapes for example, see [12,13]. Natural landscapes have been widely interpreted as a source of aesthetic, psycho-physical and social benefits in people's lives; and as a means to improve the emotional state and restore good health, see [14-17]. Thanks to the identification of the benefits provided by these goods for the general welfare of the population, the literature has focused on determining those factors which determine citizens' decision making structures and which lead them to perceive and use natural resources differently, in reference [18]. The perception of the landscape is a complex process which involves tangible issues related to vision, and psychological questions related to cognition, affect and evaluation. The perception of the resources that the natural areas have to offer and the extent to which it caters to the user's personal interests can either strengthen or weaken usage behaviour, depending on whether or not the experience is judged to be positive, in [19]. As noted by [20], the experiences felt by visitors to natural areas are not the only determining factor in their evaluation of the area or their behavioural response towards it. It is the individual's personal values that seal the decision-making process, by dictating the user's mode of action. In exploring the role of personal values, Researchers, see for example [21], have focused on human attitudes and behaviours towards natural areas, based on the theory that they derive from underlying personal values, regarded by some as the simple principles that guide evaluations or cognitive inferences. Given their role in determining people's attitudes and responses toward specific aspects of the environment, personal values need to be integrated into the analysis of decision-making processes, in [22].

The above context of analysis, which examines the individual decision structure in terms of the relationship between attitudes and personal values, provides the framework for this study, which has two objectives. The first is to determine, in a cognitive model, whether the differentiating attributes of a given environmental good provide users with benefits and reasons for use and enjoyment, later leading to the fulfillment of their personal end values; through the Means-End Chain (MEC) methodology, in reference [23]. The MEC theory assumes that the decision-makers subjective perception of a good is the result of associations between its attributes (the "means") and more abstract cognitive schemata, which include the personal values underlying certain behaviour (the "ends"). Such associations determine the appeal of the characteristics of the good in question, see reference [24]. The second is that it aims the role of environmental attitudes as a factor to explain variation across individual decision structures. Thus, the analysis tests whether this attitudinal variable discriminate the visitor's decision-making structure in terms of perceived benefits and desired values in relation to the use of a suburban park. For the implementation of these objectives, the natural area selected for this study is the Monte San Pedro Park that is located in the north-west of the Iberian peninsula (Spain).

This study aims to contribute to the existing literature in three ways. First, this work contributes to the expanded use of the MEC technique. This methodology, adopted from the field of marketing and to date rarely used in environmental studies, enabled us construct a cognitive model of how the use of a particular natural space can help people achieve a desired end state. Therefore, the results of this study will allow us to verify the validity and applicability the means–end chain method in environmental assessments. The second contribution is the relationship between visitor decision structure and their environmental attitudes, no tested at the moment in the literature. This can help to detect the influence of the environmental profile of individuals in their perceptions and evaluations of natural areas. The final understanding of the personal values that drive the green space decision-making processes may assist managers to know the factors to improve the environmental proclivity. Additionally, the analysis of landscape preferences, in terms of park users' environmental attitudes, may provide a useful tool for the design of environmental education programmms. Finally, the third contribution is the examination of ''suburban parks''. Areas of great natural value on the outskirts of metropolitan areas with a combination of natural and manmade landscapes. Suburban parks are generally neglected in the environmental literature, but they are the most frequently visited natural areas because of their location and potential for leisure activities. Thus, it is important to determine the needs and motivations of people who use and assess these areas.

2. Theoretical framework. Means-end chain theory

Knowing how a decision is linked through the cognitive structure of visitors may be of interest in environmental economics and could be of particular importance to land managers. Research on this topic has largely focused on the most concrete level of visitors' decision-making, that is, the attributes that the natural area has to offer. The intricacies of the cognitive structure, however, are such that perceptions and valuations of the attributes of the good often result in complex choice structures.

With roots in the work of Kelly [25] and developed as a tool research into human behaviour in reference [23], the Means-End Chain Theory shows the underlying reasons which justify the importance of personal values in people's decision making structures and demonstrates that there are several levels in the cognitive structure of the decision maker when making a decision for example see [26]. It is thus assumed that the decision-makers subjective perception of a good is the result of associations between its attributes (the "means") and more abstract cognitive schemata, which include the personal values underlying certain behaviour (the "ends"). Such associations determine the appeal of the characteristics of the good in question, in reference [24].

According to this theory, consumers' product knowledge is organized into hierarchical levels of abstraction, where the higher the level of abstraction, the stronger and more direct the connection to the self. Six ascending levels of abstraction describe the cognitive structure linking product knowledge (concrete attributes, abstract attributes and functional consequences) with self-knowledge (psychological consequences, instrumental values and end values), as noted by [27]. In the case in hand, the attributes are those properties or characteristics of the park, service or performance that visitors may desire or pursue. The abstract attributes are those whose verification is impossible to check prior to use except through internal or external information sources. Exploration of the knowledge structure in the environmental context clearly established the distinction between concrete and abstract attributes for example, see [28]. Functional consequences are the benefits obtained directly from the use of the park. Psychosocial consequences are of a more personal, social and less tangible nature. Instrumental values are intangible goals related with the behavioural means used to achieve the end aims and, finally, terminal values refer to desired end states, see reference [29].

Within the environmental field it is appropriate to highlight several previous studies in a context close to this one which use this methodological approach. For example, there is the work of [30] who studied social structure and social relations to determine the quantity and quality of the environmental impacts derived from economic activity, finding that the dynamic of the time structure depends on the skills of the citizens and human capital. In [31] it is used the Means-End Chain theory to study people's motivation in connection with obligatory recycling in the Netherlands, and showed that it was environmental values and values related to the duty of the citizen which led people to recycle. Prior research attempting to identify recycling goals and their effect on the decision to recycle had reached similar conclusions; see for example [32]. Finally, the paper closest to our area of study is by [33], who signaled usefulness of the Means-End Chain perspective for the management and planning of the recreational services to be found in green spaces. Thus, by way of this approach, they obtained a better understanding of the landscape's features and the consequences, values and needs of users that determine preferences for certain goods and services associated with these spaces.

Recently, Means-End Chain theory has been applied to suburban parks to determine the similarities and differences in the decision making structure of visitors, through the study of willingness to pay for the recreational use of the good analyzed or through attitudinal variables such as satisfaction, in references [34, 35]. It has thus been shown that the visitors willing to pay for the use of the parks, or those most satisfied with them, have a more complex decision making structure with regard to the benefits and personal values received in the parks than the visitors that are unwilling to pay or are less satisfied, respectively. In this study the intention is to apply Means-End Chain theory, to determine the cognitive structure of suburban park visitors based on their environmental attitudes. Following the results obtained in previous studies, we hypothesize that the stronger the pro-environmental attitudes individuals have, the greater will be the complexity of their decision making structure, as due to this positive attitude towards the environment they will receive greater benefits and values during their stay in the parks.

Initial hypothesis: Individuals with a strong pro-environmental attitude have a more complex decision making structure than individuals with weak environmental attitudes.

3. Methodology

3.1. Study area

The area selected for the study was 'Monte San Pedro Park', opened on June 6 1999, is located in the north-west of the Iberian Peninsula (Spain). It is a large leisure-oriented, extra-urban, aesthetically up-to-date, topographically-varied area measuring 7.84 ha, and offering vistas of the city of A Coruña (Galicia) and a wide strip of coast line, stretching from Cape San Adrián and the Sisargas Islands which lie to the west as far as Cape Prior and Cape Prioriño to the east. The seashore is of particular interest due to characteristic rock formations, flora and fauna. The flora is dominated by yellow gorse and pink heather interspersed with other typical coastal plants, some of them unique to the area. A great variety of small birds including warblers, goldfinches and linnets add their touch of life and colour.

Abandoned army bases have left underground shelters, barracks, lookout posts and shore batteries. The panoramic view as one descends from the park includes all the major landmarks of the city below. In addition to what nature itself has provided, there are plenty of tracks and pathways, landscaped areas, public amenities, ponds and information panels, etc. Finally, the location of this park on the coast of Cantabria overlooking the Atlantic Ocean makes this an ideal place for visitors to enrich their leisure time and enhance their enjoyment and quality of life. The Monte San Pedro Park, open to the public free of charge, is an example of first class territory management and planning, offering visitors an excellent opportunity to enjoy the views, engage in sports and other recreational activities and generally relax.

3.2. Procedure and measures

Prior to the survey, a pilot study was carried out on a sample of 30 subjects to ensure the validity and user-friendliness of the questionnaire. The pilot study was developed and administered in a series of meetings and interviews with experts and focus groups (made up of potential visitors to the areas under analysis) That help us make minor adjustments. Once the pre-test was carried out, a random sampling stratified by age and gender of visitors to the suburban park was implemented. The data was collected between April and June 2010 in 230 face to face interviews carried out with citizens of A Coruña who were visiting the Monte San Pedro Park at the time of the survey. On average respondents took 20 to 25 minutes to orally complete the questionnaire with the assistance of the interviewer. The final sample consisted of 194 usable questionnaires, 36 questionnaires were rejected. This sample size exceeds the average used in other studies applying MEC theory, which [36] report as 60 interviews due to the numerous links generated by this type of methodology. Thus the acceptance rate was 84.3%. On the basis of the final sample it can be deduced that the regular visitors, for this kind of area, are women (58.2%) aged 40 (42.9 years), who have completed secondary school (31.4%) or are graduates (35.6%) and have a medium level of income (57.7%).

A four-part questionnaire was used. Part 1 contained questions about users' attitudes and behaviours during their visit to the park. Environment-related questions made up part 2 of the questionnaire. It consists of six questions designed to elicit the user's environmental attitudinal and behavioural response. One of these questions measures the New Environmental Paradigm (NEP). Following the work of [37], the NEP scale measures *"a paradigm or worldview that influences attitudes and beliefs toward more specific environmental issues"*. In this case, the NEP scale was chosen to assess the general pro-environmental attitudes of respondents with regard to the environment and its components. Table 1 shows the average valuations offered by respondents on the NEP scale.

The third set of questions was designed for the laddering interviews, to elicit means-end chain data from respondents and thus determine the benefits expected and values pursued through use of the park. The laddering interview questions are shown in Annex. A literature revision and the pilot survey guided the choice of attributes, consequences and values considered in the survey designed to reveal park visitors' cognitive structures. It comprised eight attributes representing the concrete and abstract characteristics of the park in [18,38] and eight functional and psychological consequences in [10,14] relating to its use. The values for the analysis presented in this paper were adapted from the LOV (list of values) proposed by [39], later modified by the Rokeach Value Survey (RVS), which identifies nine key personal values that influence people's lives. The final part of the questionnaire was used to collect data to identify the socio-economic profile of the respondents.

Attitudinal measure	Scale items	Mean	SD
New Environmental Paradigm *9 items on 5-point Likert-scale. Fully disagree-fully agree*	The so-called "ecological crisis" facing humankind has been greatly exaggerated	3.07	1.11
	The balance of nature is strong enough to cope with the impacts of modern industrial nations	2.45	.91
	Humans will eventually learn enough about how nature works to be able to control it	2.65	.88
	Human ingenuity will insure that we do NOT make the earth unlivable	2.53	.89
	Humans were meant to rule over the rest of nature	2.25	.86
	Humans have the right to modify the natural environment to suit their needs	2.22	.85
	When humans interfere with nature it often produces disastrous consequences	3.32	1.09
	Plants and animals have as much right to live as humans	3.46	1.08
	Humans are severely abusing the environment	3.49	.97
	The balance of nature is very delicate and easily upset	3.55	.93
	If things continue on their present course, we will soon experience a major ecological catastrophe	3.23	.85
	We are approaching the limit of the number of people the earth can support	3.29	.99
	The earth is like a spaceship with very limited room and resources	3.31	.99
	Despite our special abilities humans are still subject to the laws of nature	3.29	.99
	The earth has plenty of natural resources if we just learn how to develop them	3.03	.93

Table 1. Complete NEP question and descriptive findings.

3.3. Data analysis

With respect to the choice of data collection method for the MEC application, the most widely known information-gathering technique is one known as "Laddering", which was first developed by [40]. Based on the personal construct theory proposed by [25], it is a face-to-face, one on-one, in-depth, semi-structured interviewing technique designed to develop a understanding of how consumers translate product attributes into meaningful associations with respect to themselves in reference [23]. In other words, its purpose is to reveal people's motives for choosing a particular good, as noted by [41]. The general laddering interview is a three-stage process. In the first stage, respondents are required to name the main attributes on which they focus when comparing and evaluating goods. The revealed key attributes are the starting point for the second stage, which is an in-depth interview, where respondents are required to explain their relevance in terms of the perceived associated consequences and personal values. Interviewers repeatedly ask respondents "Why is that important to you?" pushing them to increasing levels of abstraction (from attributes to consequences and from there to values) until they can go no further. These results in sequences of concepts or "ladders". The objective in the third stage is to plot the concepts drawn out by the laddering technique on a so-called implication matrix, in [42]. This matrix enables the construction of a Hierarchical Value Map (HVM), which is a tree diagram mapping the respondents thought process through the various levels of abstraction in the form of a graph, as noted by [24]. implication matrix, in [42]

The two possible approaches when conducting laddering interviews are soft laddering and hard laddering, in [43]. Hard laddering refers to all interview and data collection techniques in which subjects are compelled to generate or verify associations between elements within individual ladders, in sequences that reflect increasing levels of abstraction. In soft laddering, a natural and unrestricted flow of speech is encouraged during interviews, with associations between attributes-consequences-values being reconstructed subsequently during the analysis, in [44]. Hard laddering was selected for the purposes of this study because it is quicker and less expensive than soft laddering, places less pressure on the respondent and is more suitable when working with large samples (more than 50 subjects), for example see [41,43]. Concretely, the technique used in this study is the Association Pattern Technique (APT), see [42]. The APT is based on the general laddering interview technique proposed by [23] but differs in that it is more structured and uses two independent matrices: an AC matrix (attributes-consequences) and a CV matrix (consequences-values). Gutman already proposed that, for measurement purposes, the means-end chain can be conceived as a series of connected matrices. The matrices presented to the survey participants show all the attribute-consequence and consequence-value linkages shown initially in the survey. This results in two tables showing all the possible attribute-consequence and consequence-value combinations, and thus providing a dataset of binary observations, in [42].

Another methodological issue requiring consideration when working with the APT is how many links to include on the HVM in order to obtain the most meaningful results, or "Cut-

Off Level". This indicates the number of linkages registered before a connection ceases on the map, as noted by [36]. It is not easy to determine what frequency of linkages between two levels of abstraction is meaningful or significant enough to be included on the HVM. A high cut-off level (a high frequency of linkages) will give a simpler map, involving fewer connections, hence some loss of relevant information, but greater ease of interpretation. A low cut-off level (a lower frequency of linkages) will result in a complex map that will contain a large amount of information but will be more difficult to interpret. Prior research has shown various ways of determining the cut-off point for example, see [26], but, generally, the optimum cut-off point is the one that produces a HVM with the maximum amount of information and the greatest ease of interpretation. The cut-off determination method used in this study, known as "Top- Down Ranking" developed by [36] is based on the premise that a whole group of respondents will not necessarily make the same number of linkages between two levels of abstraction. It may therefore be inappropriate always to use the same cut-off point when the number of linkages between the different levels of abstraction varies. Top-down ranking enables the HVM to include only the most frequent linkages between different levels of abstraction. In other words, it selects the linkages in order of importance (the most important linkage being the one with the most cell entries).

Having designed the data collection process and selected the methods of analysis (in this case, APT and Top-down ranking), the data are processed using MecAnalyst Plus 1.0 software. This provides an ordered set of HVMs, where the first map is the simplest and easiest to interpret while showing the most important linkages. Each successive map is more complex and features a greater number of the attributes, consequences and values mentioned by each group of respondents. The process continues until the analyst decides to end it; that is, when the right cut-off point appears to have been reached, and continuation of the process would result in uninterpretable data. The advantage of this method is that it allows observation of the linkages between levels and permits between-group comparison.

4. Results

4.1. Characterization of visitors based on environmental attitudes

Based on the stated purpose of this study, which is to analyze the cognitive structure of visitors to suburban parks and to identify the role that environmental attitudes play in the variation across individual decision structures, two groups of respondents were created. One for the group whose reported "weak environmental attitudes" (attitudes less than or equal to 3); and another for those who reported "strong environmental attitudes" (higher than 3). The two groups were first characterized to reveal their socio-demographic characteristics and between-group differences using the sample of 194 visitors. The weak environmental attitudes group represents 51% of the respondents (99/194), while the strong environmental group represents 49% of the respondents (95/194). From Table 2 it can be deduced that visitors with strong environmental attitudes are mainly female, aged between 31 and 50 on average, with a university education and a middle-class based on household income. Furthermore, it can be seen that there exist differences in gender and educational level between users with weak and strong environmental attitudes. These differences show that women and people with higher education have a stronger environmental attitude by comparison with men and people with basic education, respectively. Various authors have shown that women and people with a higher level of education are more environmentally concerned, for example see [1,45].

	Group 1 Respondents with weak environmental attitudes 51%	Group 2 Respondents with strong environmental attitudes 49%	χ^2
Age			
Under 20	5.1%	3.2%	
21-30	19.2%	22.1%	1.29
31-50	41.4%	45.3%	
51-65	28.3%	23.2%	
Over 65	6.1%	6.3%	
Gender			
Male	51.5%	31.6%	7.92***
Female	48.5%	68.4%	
Income			
Low	30.3%	24.2%	0.91
Average	55.6%	60.0%	
High	14.1%	15.8%	
Level of education			
None	26.3%	16.8%	
Primary/secondary	12.1%	10.5%	8.00**
High school	35.4%	27.4%	
University Degree	26.3%	45.3%	
n=194			

Table 2. Characterization of subjects based on reported environmental attitudes.
[a] Level of statistical significance determined by Pearson's chi-square tests (**sig<0.05, ***sig<0.01).

4.2. Decision structure analysis by means-end chain theory

Hierarchical Value Maps

HVM were constructed using a Top-Down 'Ranking' Cut-Off approach [36]: a top 7 cut-off was chosen meaning that the HVM gives a representation of the seven most frequently chosen links between two levels of abstraction. Moreover, this method to determine the cut-off can be used to obtain between-group comparisons. Table 3 shows the cutoff points at each level analyzed. As can be seen, the cutoff point differs from one map to another and between the types of relationships established. Level 1 represents the most important chain for each group of respondents. The "sport-fitness" consequence-attribute chain and the "fitness-fun" consequence-value chain were the most import for the two groups of respondents. Thus, among the respondents with weak environmental attitudes, these chains were chosen by 65.7% and 63.6%, respectively, while among those with strong environmental attitudes 70.5% and 51.6%, chose these options, respectively. However, this similarity between groups in attributes, consequences, and values obtained does not continue at later levels, with notable differences appearing between them. In this case what is shown is an example of the most important chain for the two groups of respondents corresponding to level 1. All the attribute-consequence and consequence-values chains for each group of respondents up to level 7 can be seen together in Figures 1 and 2.

		Group 1 Respondents with weak environmental attitudes		Group 2 Respondents with strong environmental attitudes	
		Cut-off point	%	Cut-off point	%
Level 1	AC[a]	65	65.7	67	70.5
	CV[b]	63	63.6	49	51.6
Level 2	AC	44	44.4	49	51.6
	CV	51	51.5	48	50.5
Level 3	AC	41	41.4	47	49.5
	CV	49	49.5	43	45.3
Level 4	AC	35	35.4	45	47.4
	CV	45	45.4	42	44.2
Level 5	AC	31	31.3	40	42.1
	CV	40	40.4	40	42.1
Level 6	AC	29	29.3	36	37.9
	CV	39	39.4	39	41.1
Level 7	AC	25	25.3	35	36.8
	CV	34	34.3	37	38.9

Table 3. Cut-off points for the 7 levels of abstraction and percentage of total cases
[a] Attribute-Consequence [b] Consequence-Value

As the level increases so does the complexity of the number of attributes, consequences and values selected for each group until level 7 is reached, from which point the quantity of information becomes impossible to interpret. Thus the cutoff point for level 7 between the attributes-consequences relationship is 25 for the group with weak environmental attitude and 35 for the group with strong environmental attitudes. In the case of the consequences-values relationship, the cutoff points are to be found between 34 and 37, respectively. This shows, a priori, that the group with a strong environmental attitude achieves a higher cutoff point and so a higher level of abstraction than the groups with weak environmental attitudes.

Figures 1 and 2 show the HVM for the two groups for a cut-off level of 7. Each element in the chain (attributes, consequences and values) is shown together with the percentage of respondents who made that particular linkage. Each chain represented on the maps varies in thickness depending on the percentage of respondents who chose it. Both groups show a high frequency of linkages between the different levels on the ladders, which provide an initial indication of the relevance of the aspects upon which the subjects were asked to form attribute-consequence-value chains. That is, visitors establish a considerable number of linkages between the attributes of the park, the benefits they obtain from it, and their own personal values. Thus, confirmation is found for the implication of personal values in subjects' perception of the various attributes or differentiating features of the environmental good, depending on the desired consequences or benefits in each case. Furthermore, in the maps there are arrows with broken and continuous lines. The arrows with continuous lines represent chains formed by an attribute, a consequence and a value, while the arrows with broken lines represent segmented chains, formed by an attribute-consequence or a consequence-value.

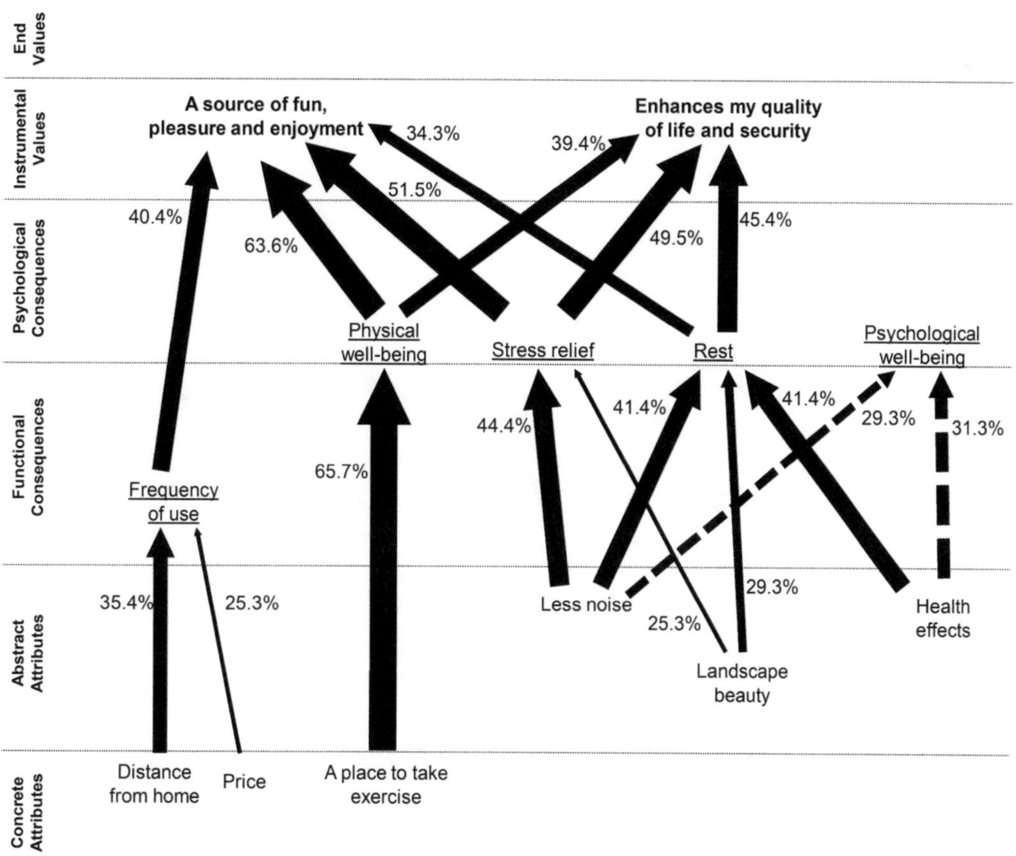

Figure 1. Hierarchical Value Map for Group 1 "Respondents with weak environmental attitudes".

An initial partial analysis of the results reveals interesting similarities between the two groups. Thus, in relation to attributes, there is a shared interest in doing sport activities, distance from home, less urban noise and landscape beauty. As far as consequences are concerned, both groups show interest in the frequency of visits and concern in physical and mental well-being. A final key aspect of the between-group comparison concerns the personal values included in the hierarchical structure, where both search values such as enjoyment and enhancing their quality of life.

On the other hand, some differences emerge between the groups. Only the group with a strong pro-environmental attitude is interested in the functional consequences "ecological habits" and "help the environment", suggesting that the environmental component arises with greater impetus among those with a more pro-environmental attitude. Furthermore, the terminal values "a sense of fulfillment" and "peace of mind and self-respect" appear only in this group. This reveals how the users who have strong environmental attitudes are more concerned with achieving personal end values, while weak-environmental attitudes users are only concerned with instrumental values. The mention of more values, and the fact that they are terminal values, suggests at first sight that those with a strong environmental attitude reach a higher level of abstraction. These initial results should be analyzed in more depth, however, and the ladders generated by the MEC application should be examined further in order to draw conclusions regarding the different levels of abstraction reached by the two groups.

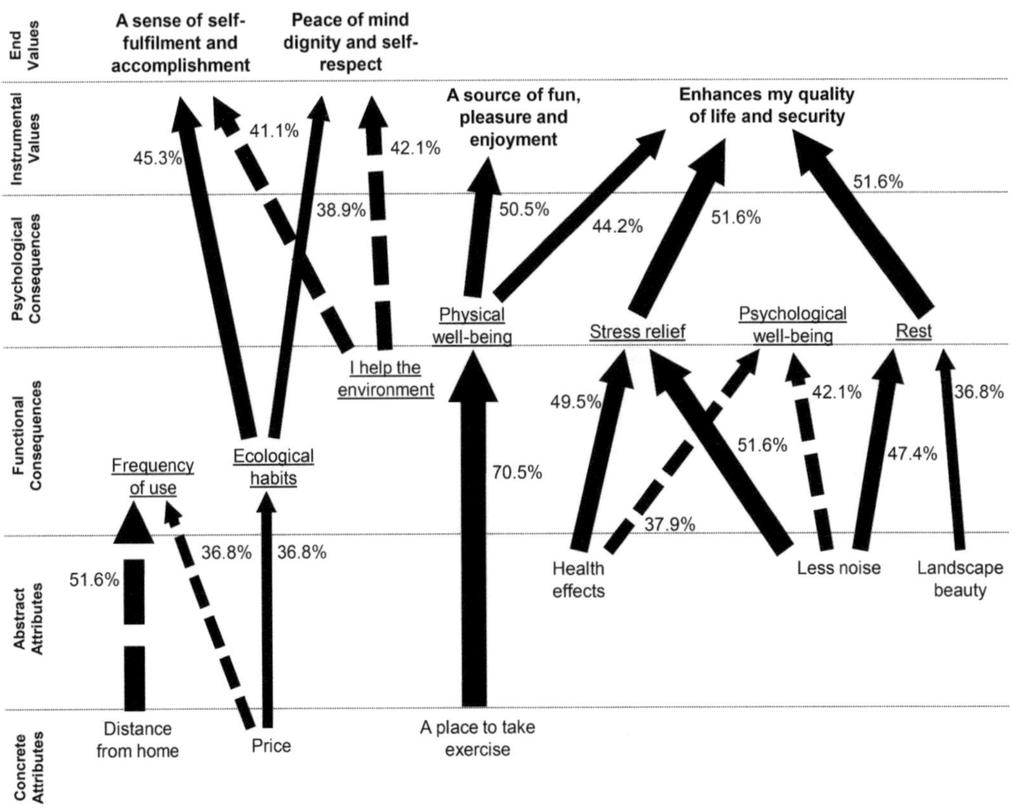

Figure 2. Hierarchical Value Map for Group 2 "Respondents with strong environmental attitudes".

Analysis of the HVMs ladders

The results described above are eligible for further analysis aimed at advancing in the understanding of the means-end chain formation process through which users' link park attributes to consequences and thence to values. Among the ladders observed in this study, it is possible to discern that four linkages are shared by all respondents, irrespective of their environmental attitudes. The first and foremost is the concrete attribute "a place to practice sport and take physical exercise" which is linked to the psychological consequence "physical well-being" (formed by 65.7% of the weak-environmental-attitude group and 70.5% of the strong-environmental-attitude group) and the instrumental values "a source of fun, pleasure and enjoyment" (63.6% versus 50.5%) and "enhances my quality of life" (39.4% versus 44.2%). This appears to suggest that one of the main values users hope to fulfill through using a suburban park is, as might be expected, the enjoyment that comes from doing exercise and improving physical fitness. Secondly, another linkage common to both groups is the one between the abstract attribute "less noise" and the psychological consequences "stress relief" (44.4% versus 51.6%) and "rest" (41.4% versus 47.4%) and the instrumental value "enhances my quality of life and security" (49.5% versus 51.6%) and (45.4% versus 51.6%), respectively. Thirdly, the abstract attribute "landscape beauty" is linked to the psychological consequence "rest" (29.3% versus 36.8%) and the instrumental value "enhances my quality of life and security" (45.4% versus 51.6%). It would therefore be interesting to improve the aesthetic appearance of the spaces with the aim of improving the visitor's emotional state and, consequently, increase the visitors' welfare and the use placed on the environmental good in question. Finally, both groups share an interest in the distance from the place of residence to the suburban park (35.4% versus 51.6%) and the cost of access to the park (25.3% versus 36.8%), which influences the frequent use of this type of area.

With regard to the differences between the two groups, only those with a strong pro-environmental attitude did not condition the enjoyment of some ecological habits to the price of access to the area (36.8%). Furthermore, this ecological attitude had a positive influence on their conscience and personal respect (38.9%) and their desire for personal growth (45.3%). At the same time, the incomplete chain which links environmental help with feelings of personal growth (41.1%) and personal conscience is of particular importance (42.1%).

Comparison of abstraction levels

With the attributes, consequences and personal values pursued by each group in accordance with the environmental attitude of the respondent, and the chains between them, established, we go on to determine the degree of abstraction attained by each of the groups. This is determined by the average number of ladders and the average number of elements (attributes, consequences and values) of each level, in order to determine whether there were differences between groups based on these criteria.

The total number of ladders constructed by the groups of visitors was 2382, of which 986 corresponded to the weak-environmental-attitude group and 1396 to the strong-environmental-attitude group. Breaking down the number of ladders proposed by each of the groups, statistically significant differences in the average of incomplete (F = 6,019, sig <0.05) and complete (F = 8,692,sig <0.01) ladders can be observed. Thus, the group with a strong environmental attitude has an average of 14.69 complete ladders and 9.96 incomplete ones, with the other group having an average of 9.96 complete ladders and 3.03 incomplete ones. These results already begin to indicate the presence of a greater degree of abstraction in the decision making structure of the groups with a strong pro-environmental attitude, as they establish associations between attributes, consequences and values.

To complete the analysis of the abstraction, Table 4 summarizes the average number of attributes, consequences and values provided by each group. Statistically significant differences were observed in the presence of more concrete and abstract attributes, functional consequences and terminal values among respondents who have a strong environmental attitude. These results indicate that respondents who have stronger environmental awareness have higher complexity or abstraction in their decision making structure, as they incorporate more aspects of their personality.

	F Snedecor	Group 1 Respondents with weak environmental attitudes	Group 2 Respondents with strong environmental attitudes
Concrete attributes	7.762***	1.94	2.44
Abstract attributes	3.903**	2.64	3.01
Functional consequences	5.028**	2.34	2.82
Psychological consequences	1.867	3.22	3.45
Instrumental values	2.560	1.85	2.08
Terminal values	11.419***	1.49	2.35

Table 4. Average numbers of attributes, consequences and values used by each group.
***sig<0.01

Complexity indices

Finally, the complexity indices developed by [46] were estimated in order to determine which group showed the more complex means-end chains in the HVMs. Two complexity estimates were calculated. The first, labeled C1, measures the complexity of the maps in terms of the concepts used. It is obtained by dividing the number of attribute-consequence-value chains by the total number of attributes, consequences and values used in the maps. The other, labeled, C2, measures the complexity of the maps in terms of the connections between ladders. It is obtained by dividing the total length of the chains by the total number of individual connections.

As can be seen from Table 5, the C_1 index is verified (index value 187 for strong environmental attitudes and 174.7 for weaker). The results show that the maps of the groups of respondents with a strong environmental attitude are more complex in terms of the number of attributes, consequences and values received during the stay. On the other hand, it has not been demonstrated with this test that this group also has greater complexity in the relationships established between the elements of the chain. Therefore, to explain the "complexity" that exists in the decision making process of individuals requires further analysis.

	Number of Cognitions (a)	Number of Links (b)	Number of Paths (c)	Total lengh of Paths (d)	C_1 (b/a)	C_2 (d/c)
Group 1 Respondents with weak environmental attitudes	13	2272	17	30	174.7	1.76
Group 2 Respondents with strong environmental attitudes	17	3179	18	22	187	1.22

Table 5. Complexity indices.

5. Discussion

This study differs from the existing research in two main aspects. The first is the use of Means-End Chain methodology in a cognitive model to determine whether the differentiation attributes of a given natural area are associated with benefits and reasons for use that ultimately relate to individuals' terminal values or personal values. The second aspect that is worthy of note has to do with differences in perception and the previous decision making structure of the individual depending on his or attitude to the environment and its components (whether the individual has a strong pro-environmental attitude or not), in terms of the benefits expected and the personal values that he or she wishes to achieve.

To achieve these ends the Means-End Chain methodology was used by way of a laddering interview. The observation of the Hierarchical Value Maps leads to two types of results. Firstly, there are the results shared between the two groups analyzed. Thus, we see that both respondents who have a strong pro-environmental attitude as well as those with a weak environmental attitude shared an interest in tangible characteristics such as sports, the price of access to the park and the distance from the residence of the visitor. This finding confirms the results of previous studies [47,48]. Among the most valued intangible characteristics in the study are effects on health, noise reduction and the beauty of the landscape. Numerous authors, for example see [17,49], have shown that the search for beautiful landscapes and the restorative and therapeutic power offered by nature are highly valued factors. As for the benefits received, there is special emphasis on the physical and mental wellbeing produced by being in these areas. Interest in these benefits was also found by researchers in other environmental contexts, in [14,50,51]. Finally, there is also the shared interest in various personal values, individual in nature, such as entertainment and improvement of the quality of life. In general terms the relevance of personal factors in the valuation of these natural areas was confirmed in [15,52,53]. Therefore, we may affirm that the attributes, consequences and values mentioned are appreciated and sought after by all kinds of people when they use and place a valuation on suburban parks, points to note for land managers in their management strategies.

Secondly, there is another type of result, concerning the analysis of the differences obtained from the cognitive structures of the interviewees on the basis of their environmental attitudes. The presence of different benefits and values suggests that the perceptions of the users vary in accordance with the degree of their environmental pro-activeness. Thus the degree of involvement in environmental and ecological questions as well as consciousness raising and personal development of those with a strong pro-environmental attitude is noteworthy. That is to say, the groups with a strong environmental attitude show a greater level of ecological awareness and are more interested in achieving terminal type personal values. It is thus shown that environmentally conscious individuals have a more significant emotional dimension, i.e. they have a greater level of abstraction in their decision-making structures when it comes to using and placing a value on natural areas, so confirming our initial hypothesis. The environmental literature to date has focused on various socio-economic and socio-demographic factors that determine the environmental attitudes of individuals and how these attitudes condition the behavioural decisions of individuals, with regard to environmental conduct, for example [1-3,8-11,54] The present study shows, for the first time, how environmental attitudes determine individuals' decision making structures. This information may be of use to land managers as ecological issues have been shown to be a differentiating element in the decision-making structure of visitors to suburban parks.

6. Conclusions

On the basis of these results we can conclude that people are emotionally linked to the environment in terms of their preferences and the personal values satisfied during their stay. The possibility of doing sports, distance from place of residence, health benefits received and the search for fun and to improve quality of life have emerged as key factors in the use and valuation of suburban parks. For this reason, planning strategies could include the provision of sport and relaxing landscapes to improve the physical and emotional state of the visitors and their recreational and well-being opportunities. In the context of the current urban reality, where there are multiple leisure opportunities, it is necessary to conduct studies of the population near the location of urban and suburban and natural areas to help managers to become acquainted with the decision-making process of these citizens. As these results have demonstrated, knowledge of citizens' psychology is indispensable for ensuring that the provision of services in natural areas is in accordance with the expectations, desire and necessities of their potential visitors. The satisfying of these desires and necessities could lead to positive behaviour from citizens such as, for example, greater use, global valuation and conservation of these places are so valuable for human existence.

Furthermore, individuals with a strong attitude in favor of the environment have a more pronounced emotional dimension, which manifests itself in the fact that they have a greater degree of complexity in their perception of benefits and values during their stays in urban parks. Thus, interest in ecological practices and values relating to personal conscience and respect, are only seen in this group of visitors. For this reason, managers should also seek to strengthen the pro-environmental attitudes and beliefs of individuals in order to differentiate suburban parks on the basis of their uses and the necessities of individuals. For managers to achieve these objectives, the main strategy employed must be environmental education. The design of environmental education programs (e.g. courses, conferences and/or talks) may increase environmental awareness through encouraging introspection in the community into environmental attitudes and ethical issues that should be promoted in order to induce changes in the behaviour of individuals. As noted by this study, attitudes determine visitors' decision-making structure and, consequently, affect their perception and decisions regarding behaviour.

Other interesting conclusions may include the fact that the greater the benefits received from environmental goods, the greater is the involvement of the person concerned with these goods. It would therefore be interesting to communicate to the whole population the benefits, both physical and psychological, that are freely offered to us by the environment in order to get them more involved in pro-conservation attitudes, in respect and protection of the environment. Of course the receiving of health and psychological benefits has an effect on the social cost of healthcare, which is another positive aspect to be considered about these spaces. Furthermore, the closeness of suburban parks allows easy access to them by urban residents, and a resulting greater capacity to transmit the benefits provided by the presence of these goods. These factors could be taken into account by land managers who could include the health benefits produces by these natural resources in their calculations and use them in decision making when the relationship between their costs and benefits arise. It would thus be very interesting for experimental designs to be jointly carried out with the managers of these areas, and to the degree that it might be possible, representatives of the local population as well.

Finally, the results of this study have confirmed the validity of the Means-End Chain methodology in the valuation of the environment. This technique allows a more precise definition of the user experience of the visitor. Thus, it identifies those elements that are unique to the park, which serves as a bridge to understand the underlying motives in the minds of visitors and determine their environmental preferences. The greater the success of managers in identifying Means-End Chains, the greater will be their capacity to help visitors to achieve their leisure goals, which will have a positive effect on their current and future use of this environmental good. It has also been shown that Hierarchical Value Maps provide a basis for effectively communicating the search for individual or social values by potential park visitors. These maps thus hierarchically link the attributes, consequences and values that define the way these individuals see themselves, their place in the world and later, how they develop these preferences in certain types of experiences. From all of which it can be concluded that Hierarchical Value Maps provide sufficient information on which to base research.

Future researchers may find it interesting to employ other methodological perspectives or social psychological theories to examine the discriminating power of the environmental attitudes of individuals in order to better calibrate its capacity to explain the decisions of users of the environment, or regarding the decision not to make use of the environment. It would also be interesting to learn about the cognitive structures of people who do not make use of these spaces, in order to establish differences between users and non-users. Finally, other attributes and benefits of natural spaces could be identified as well as other personal values involved, which might influence the economic valuation of this type of areas.

Author details

Natalia López-Mosquera, Mercedes Sánchez and Ramo Barrena
Public University of Navarra, Business Management Department, Pamplona, Spain

Appendix

Annex. Laddering interview

Now, using the tables below, we would like you to identify, first, the relationships between the characteristics of the San Pedro Park and the consequences of your visit as you see them, and then the relationships between those consequences and the benefits you wish to obtain from your visit. In both tables you are asked to indicate linkages between the rows and the columns. The idea in the first table is to tick a box (or boxes) to indicate the link(s) between each characteristic (a1, a2, ..., a13) and the benefit(s) which, in your experience, it provides (c1, c2, ..., c22). In the second table, you are asked to do the same to link each benefit in the columns (c1, c2, ..., c22) with the personal value(s) to which, from your experience, you think it caters (v1, v2, ..., v9).

		a1 Price	a2 Landscape beauty	a3 A place to practice sport and take physical exercise	a4 Distance from home	a5 Closeness to nature	a6 Health effects (oxigenation, air purification)	a7 Recreational facilities provided	a8 Less noise
c1	I have good ecological habits								
c2	Stress relief /relaxation								
c3	Physical well-being /improved fitness								
c4	Rest								
c5	I help the environment								
c6	A way to escape from routine								
c7	Frequency of use /regular user of green space								
c8	Mental well-being								

Table 6. Attributes-Consequences. Link attributes (a1, a2,...) with consequences (c1, c2, c3...)

		v1 A sense of social belonging	v2 A source of fun, pleasure and enjoyment	v3 Better relationships with others	v4 Enhances my quality of life and security	v5 A sense of self-fulfilment and accomplishment	v6 Emotional stimulation	v7 Being respected by others	v8 Peace of mind dignity and self-respect	v9 I'm more successful
c1	I have good ecological habits									
c2	Stress relief /relaxation									
c3	Physical well-being /improved fitness									
c4	Rest									
c5	I help the environment									
c6	A way to escape from routine									
c7	Frequency of use /regular user of green space									
c8	Mental well-being									

Table 7. Consequences-Values. Link consequences (c1, c2, c3,…) with values (v1, v2, v3…)

7. References

[1] Torgler B, Garcia-Valiñas M.A. The Determinants of Individuals' Attitudes Towards Preventing Environmental Damage. Ecological Economics 2007; 63 536-552.

[2] Witzke H.P, Urfei G. Willingness to pay for environmental protection in Germany: coping with the regional dimension. Regional Studies 2001; 35(3) 207–214.

[3] Veisten K, Hoen H.F, Navrud S., Strand, J. Scope insensitivity in contingent valuation of complex environmental amenities. Journal of Environmental Management 2004; 73 317–331.

[4] Ajzen I. The theory of planned behavior. Organizational Behavior and Human Decision Processes 1991; 50 179–211.

[5] Stern P.C, Dietz T, Abel T, Guagnano G.A, Kalof L. A value-belief-norm theory of support for social movements, the case of environmental concern. Human ecology review 1999; 6(2) 81-97.

[6] Ford R.M, Williams K.J.H, Bishop I.D, Webb T. A value basis for the social acceptability of clearfelling in Tasmania, Australia. Landscape and Urban Planning 2009; 90 196-206.

[7] Wauters E, Bielders C, Poesen J, Govers G, Mathijs E. Adoption of soil conservation practices in Belgium: an examination of the theory of planned behaviour in the agri-environmental domain. Land Use Policy 2010; 27 86–94.

[8] Oreg S, Katz-Gerro T. Predicting Proenvironmental Behavior Cross-Nationally: Values, the Theory of Planned Behavior, and Value-Belief-Norm Theory. Environment & Behavior 2006; 38 462-483.

[9] Fielding K.S, McDonald R, Louis W.R. Theory of planned behaviour, identity and intentions to engage in environmental activism. Journal of Environmental Psychology 2008: 28 318-326.

[10] Bernarth K, Roschewitz A. Recreational benefits of urban forests: explaining visitors` willingness to pay in the context of the theory of planned behavior. Journal of Environmental Management 2008; 89 155-166.
[11] Spash C, Urama K, Burton R, Kenyon W, Shannon P, Hill G. Motives behind willingness to pay for improving biodiversity in water ecosystems: Economics, ethics and social psychology. Ecological Economics 2009; 68 955-964.
[12] Winter C, Lockwood M. A model for measuring natural area values and park preferences. Environmental Conservation 2005; 1334(32) 1–9.
[13] Golicnik B, Ward C. Emerging relationships between design and use of urban park spaces. Landscape and Urban Planning 2010; 94 38-53.
[14] Chiesura A. The role of urban parks for the sustainable city. Landscape and Urban Planning 2004; 68 129-138.
[15] Tyrväinen L, Mäikinen K, Schipperijn J. Tools for mapping social values of urban woodlands and other areas. Landscape and Urban Planning 2007; 79 5-19.
[16] Velarde M.D, Fry G, Tveit M. Health effects of viewing landscapes- landscape types in environmental psychology. Urban forestry and Urban Greening 2007; 6 199-212.
[17] Korpela K.M, Ylén M, Tyrväinen L, Silvennoinen H. Determinants of restorative experiences in everyday favorite places. Health and Place 2008; 14 636-652.
[18] Krenichyn K. The only place to go and be in the city: women talk about exercise, being outdoors, and the meaning of a large urban park. Health and Place 2006; 12 631-643.
[19] Driver B.L, Brown Perry J. A socio-psychological definition of recreation demand, with implications for recreation resource planning. Assessing demand for outdoor recreation. National Academy of Sciences, Washington, D.C.; 1975.
[20] González A. La preocupación por la calidad del medio ambiente. Un modelo cognitivo sobre la conducta ecológica. Tesis doctoral Universidad Complutense de Madrid; 2002.
[21] Rokeach M. The nature of human values. Free Press, New York; 1973.
[22] Perugini M, Bagozzi R.P. The role of desires and anticipated emotions in goal-directed behaviours: Broadening and deepening the theory of planned behavior. British Journal of Social Psychology 2001; 40 79-98.
[23] Gutman J. A Means-End Chain Model Based on Consumer Categorization Processes. Journal of Marketing 1982; 46 60-72.
[24] Reynolds T.J, Gutman J. Laddering theory, Method, Analysis and Interpretation. Journal of Advertising Research 1988; 28 11-31.
[25] Kelly G.A. The psychology of personal constructs. New York: Norton; 1955.
[26] Pieters R, Baumgartner H, Allen DA means-end chain approach to consumer goal structures. International Journal of Research in Marketing 1995; 12 227-244.
[27] Olson J.C, Reynolds T.J. Understanding Consumers' Cognitive Structure: Implications for Advertising Strategy. In L. Percy and A. Woodside (Eds.), Advertising and Consumer Psychology. Lexington, MA: Lexington Books; 1983.
[28] Purcell A.T.. Abstract and specific physical attributes and the experience of landscape. Journal of Environmental Management 1992; 34 159–177.
[29] Miele M, Parisi V. Consumer concerns about animal welfare and food choice. Italian report on laddering interviews. March; 2000.

[30] Cogoy M. The consumer as a social and environmental actor. Ecological Economics 1999; 28 385-398.

[31] Smeesters D, Warlop L, Cornelissen G, Vanden Abeele P. Consumer motivation to recycle when recycling is mandatory: two explanatory studies. Journal of Economy and Management 2003; 48(3) 451-468.

[32] Bagozzi R, Dabholkar P. Consumer recycling goals and their effect on decisions to recycle: A mean end chain analysis. Psychology and Marketing 1994; 11 1-28.

[33] Frauman E, Cunningham P.H. Using a means-end approach to understand the factors that influence greenway use. Journal of Park and Recreation Administration 2001; 19(3) 93-113.

[34] López-Mosquera N, Sánchez M. The influence of personal values in the economic-use valuation of peri-urban green spaces: An application of the means-end chain theory. Tourism Management 2011; 32 875-889.

[35] López-Mosquera N, Sánchez M. The Role of satisfaction and emotional response in the choice mechanisms of suburban natural-areas users. Environmental Management 2012; 49 174-191.

[36] Leppard P, Russel C.G, Cox D.N. Improving Cadena-Medio-Fin studies by using a ranking method to construct hierarchical value map. Food Quality and Preference 2004; 15 489-497.

[37] Dunlap R.E, Van Liere K.D, Mertig A.G, Jones R.E. Measuring endorsement of the new ecological paradigm: a revised NEP scale. Journal of Social Issues 2000; 56(3) 425– 442.

[38] Gobster P, Westphal L. The human dimensions of urban greenways: planning for recreation and related experiences. Landscape and Urban Planning 2004; 68 147-165.

[39] Kalhe L.R. The nine nations of North America and the value basis of geographic segmentation. Journal of Marketing 1985; 50 37-47.

[40] Hinkle D.N. The change of personal constructs from the viewpoint of a theory of construct implications. Unpublished doctoral dissertation, Ohio State University, Columbus, 1965.

[41] Russell C.G, Flight I, Leppard P, Van Lawick V.P, Syrette J.A, Cox D.N. A comparison of paper-and-pencil and computerized methods of hard laddering. Food Quality and Preference 2004; 15 279-291.

[42] Ter Hofstede F, Audenaert A, Steenkamp J-B.E.M, Wedel M. An investigation into the association pattern technique as a quantitative approach to measuring means-end chain. International Journal of Research in Marketing 1998; 15 37-50.

[43] Grunert KG, Grunert SC. Measuring subjective meaning structures by the laddering method. Theoretical considerations and methodological problems. International Journal of Research in Marketing 1995; 12 209-225

[44] Costa A.I.A, Dekker M, Jongen W.M.F.. An overview of means-end theory: potential application in consumer-oriented food product design. Trends in Food Science and Technology 2004; 15 403-415

[45] Arnberger A, Eder R, Allex B, Sterl P, Burns R. Relationships between national-park affinity and attitudes towards protected area management of visitors to the Gesaeuse National Park, Austria. Forest Policy and Economics 2012; 19 48–55.

[46] Bagozzi P.R, Dabholkar, P.A. Discursive psychology: an alternative conceptual foundation to means-end chain theory. Psychology & Marketing 2000; 17(7) 535-550.

[47] Neuvomen M, Sievänen T, Tönnes S, Koskela T. Access to green areas and the frequency of visitis- A case study in Helsinki. Urban forestry and Urban Greening 2007; 6 235-237.

[48] Schipperijn J, Ekholm O, Stigsdotter U, Toftager M, Benstsen P, Kamper-Jorgensen F, Randrup T. Factors influencing the use of green space: Results from a Danish national representative survey. Landscape and Urban Planning 2010; 95(3) 130-137.

[49] Brown G, Raymond C. The relationship between place attachment and landscape values: Toward mapping place attachment. Applied Geography 2007; 27 89-111.

[50] Gidlöf-Gunnarsson A, Öhrström E. Noise and well-being in urban residential environments: The potential role of perceived availability to nearby green areas. Landscape and Urban Planning 2007; 83(2-3) 115-126.

[51] Lafortteza, R, Carrus G, Sanesi G, Davies C. Benefits and well-being perceived by people visiting green space in periods of heat stress. Urban Forestry and Urban Greening 2009; 8 97-108.

[52] Mill G.A, Van Rensburg T.M, Hynes S, Dooley C. Preferences for multiple use forest management in Ireland: citizen and consumer perspectives. Ecological Economics 2007; 60 642-653.

[53] Hanley N, Ready R, Colombo S, Watson F, Stewart M, Bergmann E.A. The impacts of knowledge of the past experiences for future landscape change. Journal of Environmental Management 2009; 90 1404-1412.

[54] Sauer U, Fischer A. Willingness to pay, attitudes and fundamental values- On the cognitive context of public preferences for diversity in agricultural landscapes. Ecological Economics 2010; 70 1-9.

Chapter 7

Evaluating Management Effectiveness of National Parks as a Contribution to Good Governance and Social Learning

Michael Getzner, Michael Jungmeier and Bernd Pfleger

Additional information is available at the end of the chapter

http://dx.doi.org/10.5772/50092

1. Introduction

From the perspective of sustainability, protected areas (PA) should serve several aims such as biodiversity conservation, regional economic development, social inclusion and sharing of benefits from conservation as put forward by the Convention in Biological Diversity (CBD; see Secretariat of the CBD, 2005). As such, protected areas certainly contribute to a sustainable development from various perspectives (Getzner and Jungmeier, 2012). However, as several analyses show, many protected areas can be considered as "paper parks" (Brandon et al., 1998; Bruner et al., 2001) for which no effective regulatory and management system is in place.[1] "Paper parks" not only appear in developing countries (cf. Bonham et al., 2009) but are also regular in industrialized high-income countries. For instance, the European Union's Natura 2000 network of protected areas according to the FFH (flora-fauna-habitat directive; European Council, 1992) has been estimated to require at least about EUR 5.8bn per year to be efficiently and effectively managed to fulfill its aims and objectives (Gantioler et al., 2010). Incomplete implementation frameworks, "bad governance", and the lack of resources are most prominent in Natura 2000 regions, especially in Central and Eastern Europe (cf. Kirchmeir et al., 2012).

For fulfilling the aims and objectives of protected areas, it is of great importance that property rights are well defined, the legal, institutional, and managerial frameworks are in place, and sufficient resources are provided for funding efficient and effective protected area management bodies. In comparison to the status quo, even some basic funding might significantly increase biodiversity conservation in "paper parks" (Joppa et al., 2008).

[1] The ineffectiveness of "paper parks" may also constitute a reason why studies also found no significant difference in management effectiveness of land inside and outside of parks (Hayes, 2006).

Evaluation and monitoring tools generally provide the basis for assessing these frameworks both in terms of efficiency (e.g. wise use of resources relative to outputs and results), effectiveness (e.g. achievement of ecological objectives), and social and distribution issues (e.g. benefit sharing) (see Pomeroy et al., 2005) with the aim to improving adaptive multi-dimensional management. However, merely assessing management effectiveness by applying evaluation and/or monitoring tools, while necessary, might not be sufficient for achieving the protected area's objectives. For increasing PA management effectiveness, the involvement and the contributions of several stakeholder groups might be crucial. For instance, involving even visitors and local residents can contribute to the effectiveness of ecological management since illegal behavior is reduced by increasing the understanding of and support for ecosystem management (Powell and Vagias, 2010). In this context, evaluating management effectiveness in protected areas might not only serve as an information and management tool for the PA management authority but also as an instrument for informing stakeholders and collecting tacit knowledge. Such instruments and methods thus also may contribute to "good governance" of protected areas, and lead to social learning experiences of all stakeholders. Some broad acceptance of stakeholders is thus considered to constitute an important pillar for effective (adaptive and integrative) PA management.

The current chapter therefore aims at discussing the role and function of evaluating management effectiveness as an integral part of the whole set of management tools and strategies of the PA management authority, as an important ingredient of the governance system of protected areas, and as a tool for social learning and inclusion. As a case study, we use an evaluation tool that has formerly been developed by the Nature Conservancy (2004) (Parks in Peril Site Consolidation Scorecard). This tool was transformed and adapted to the European context by Pfleger (2007a and 2008) and renamed as "European Site Consolidation Scorecard" (ESCS, Pfleger, 2007b). The most important characteristics, described in more detail below, are the strategic and long-term assessment of relevant management plans, instruments and resources, the selective inclusion of several stakeholder groups internal and external to the protected area, and the built-in feed-back and commentary sections in the evaluation report. As an example for applying the ESCS to a concrete protected area, we chose the Gesäuse National Park (Austria), established in 2002 and evaluated in 2007/2008.

The structure of the chapter is the following: Section 2 presents a brief review of the international debate concerning the application of diverse PA management effectiveness evaluation frameworks, and discusses selected aspects of evaluation in PAs. Section 3 describes the linkages between the ecological system, the economy, PA management, and the function of evaluations such as the ESCS. Section 4 presents the effectiveness indicators of the European Site Consolidation Scorecard (ESCS) and discusses the dimensions of evaluation along the empirical case study of the Gesäuse National Park (Austria). Section 5 finally discusses the importance of evaluation instruments, specifically the ESCS, for good governance and social learning, summarizes the results, and concludes with a range of further recommendations for improving the existing evaluation and monitoring frameworks.

2. Overview of frameworks for evaluating management effectiveness in protected areas

Since about the last 20 years, a broad range of frameworks (tools) for measuring management effectiveness of protected areas have been developed. Hockings et al. (2006, 1) define the evaluation of management effectiveness as "the assessment of how well protected areas are being managed – primarily the extent to which management is protecting values and achieving goals and objectives."[2] Evaluation management effectiveness of protected areas is thus discussed by Hockings et al. (2006, 1) in three dimensions;

- "design issues relating to both individual protected areas and protected area systems;
- adequacy and appropriateness of management systems and processes; and
- delivery of protected area objectives including the conservation of value."

An ideal assessment methodology and framework tries to account for the whole evaluation/management cycle (PAME, Protected Areas Management Effectiveness) with the elements of "context", "planning", "inputs", "processes", "outputs", and "outcomes".

Effectiveness of PA management in conserving biodiversity thus addresses basically the following two questions (Chape et al., 2005, 443):

- "Effectiveness of coverage: how much and what biodiversity is included within protected areas?
- Effectiveness in achieving conservation objectives: are protected areas being managed effectively?"

Leverington et al. (2010a) describe about 40 different approaches to evaluation, ranging from rapid assessments to in-depth monitoring tools. A most recent overview of management effectiveness assessment tools lists about 50 different methodologies (WDPA, 2012), of which the most widely internationally used are the following (Leverington et al., 2010b; Stoll-Kleemann, 2010; see Table 1):

- RAPPAM (Rapid Assessment and Prioritization of Protected Area Management) was developed by WWF and has widely been used for a quick assessment of strengths and weaknesses of PA management, and PA systems and networks. By allowing for comparisons between protected areas, the methodology can also be used for setting policy priorities both of governments and NGOs.
- METT (Management Effectiveness Tracking Tool) was developed to quickly assess management effectiveness at the site level using scorecards. METT is internationally used by WWF, the World Bank, and GEF (Global Environment Facility).
- The US-based Nature Conservancy (TNC) developed a specific assessment tool for its PiP (Parks in Peril) program, the Parks in Peril Site Consolidation Scorecard (The Nature Conservancy, 2004), to explore the existing and needed resources for effective management of parks that may only exist "on paper".

---

[2] Evaluations in general may fulfill several aims and objectives in modern societies. It is not only necessary to evaluate products (outputs) but to assess also outcomes in terms of legitimacy of public activities, and to provide steering mechanisms for public decision-makers (cf. Stockmann, 2007; Diller, 2008).

The tools provided or promoted by big international institutions have become standards even if some other solutions may also have their advantages or justifications (e.g. IPAM-toolbox [www.ipam.info], or the ESCS as presented in this paper).

The European Site Consolidation Scorecard (ESCS) was developed on the basis of the TNC-PiP assessment methodology and may be described as a comprehensive assessment of a specific park and its long-term policies by means of a range of indicators measured and described along a five-point scorecard (Pfleger, 2007b and 2010). Table 2 presents an overview of the indicators of the ESCS to be measured during the evaluation process.

Methodology	Abbreviation	No. of assessment with the methodology
Rapid Assessment and Prioritization of Protected Area Management	RAPPAM	939
Management Effectiveness Tracking Tool	METT	865
New South Wales State of Parks (Australia)	NSW SOP	639
Monitoring Important Bird Areas	BirdLife	506
PROARCA/CAPAS scorecard evaluation	PROARCA/CAPAS	483
TNC Parks in Peril Site Consolidation Scorecard	PiP SCSC	300
Victoria's State of Parks	Victorian SOP	102
AEMAPPS: PAME with Social Participation-Colombia	AEMAPPS	89
Parks profiles	Parks profiles	62

Source: Authors' own compilation based on Leverington et al., 2010a.

Table 1. Recent assessments of protected areas by the most prominent management effectiveness evaluation tools

No.	*Indicator*
A.	**Strategic planning**
A.1	Project area zoning
A.2	Site-based long-term management plan
A.3	Science and information needs assessment for project area
A.4	Monitoring plan development and implementation for project area
B.	**Basic protection activities**
B.1	Physical infrastructure for project area
B.2	On-site personnel
B.3	Training Plan for On-site Personnel
B.4	Land tenure and land use issues within the project area
B.5	Threats analysis for the project area
B.6	Official declaration of protected area status for the project area
B.7	Organisational structure
C.	**Long-term financing**
C.1	Long-term financial plan for sites in the project area
D.	**Site constituency**
D.1	Broad-based management committee/technical advisory committee for project area
D.2	Institutional Leadership for the project area
D.3	Common Leadership for the project area
D.4	Community involvement in compatible resource use at the project area
D.5	Stakeholder and Constituency Support for Project Area
D.6	Policy agenda development at national/regional/local levels for project area
D.7	Communication plans for the project area
D.8	Environmental education plans for the project area
D.9	Cooperation with other organizations
D.10	Integration in an ecological network

Source: Pfleger, 2007b, 81.

Table 2. Summary of scorecard indicators of the European Site Consolidation Scorecard (ESCS)

Deciding on the "right" evaluation tool, of course, has to be done in close reference to the aims of the evaluation, and also to the efforts necessary to achieve the evaluation objectives. Comparing different evaluation methods may thus be done according to the benefit-cost ratio of such instruments. Figure *1* presents a rough picture of the benefits of the ESCS in relation to the efforts (costs) involved in comparison to two prominent evaluation tools mentioned above (RAPPAM, METT). While RAPPAM is considered to be a fast evaluation providing a quick picture of PA effectiveness, METT goes a little bit more into detail. The ESCS is an evaluation tool going into much more detail. However, depending on the requirements of a specific evaluation (e.g. self-assessment vs. in-depth external evaluation), its application range is very flexible and so are the necessary efforts.

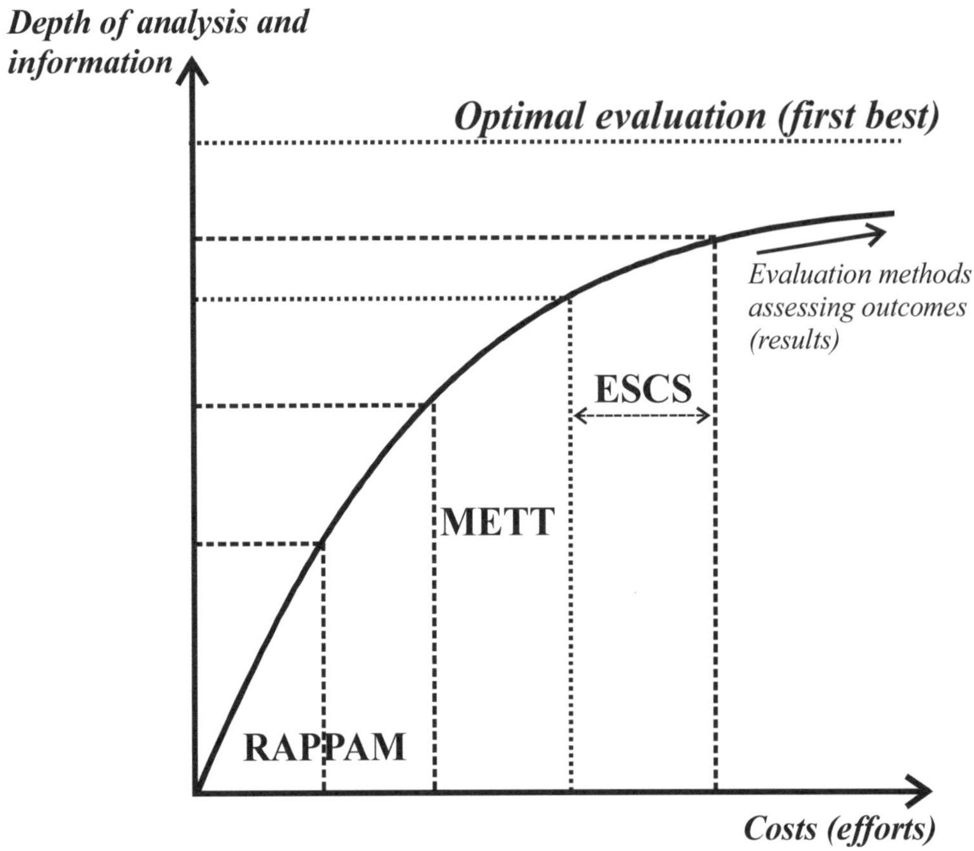

Source: Authors' draft based on Pfleger et al. (2009).

Figure 1. Cost-benefit ratios of different evaluation tools

Management effectiveness evaluation is generally considered to constitute an important ingredient and pre-requisite for adaptive management since the changing ecological, social and economic environments call for flexible management frameworks. Effective management in this sense has to focus on adaption and learning in order to hold the developing PA on a sustainable path. Thus, evaluation not only measures results (outputs) but relates outcomes to resource inputs (e.g. budgets) and also may form the basis of a monitoring system. In addition, management effectiveness instruments provide an overview of different dimensions of protected areas (ecological, social) which are not considered in mere economic valuation methods. Figure 2 presents the "management effectiveness triangle". Throughout the phases of management (the life-cycle of the protected area), certain measures and policies can be implemented. The ESCS concentrates on contexts to processes (the first four phases of management) since these are the most important to achieve respective outputs and outcomes (results). The figure also demonstrates that the decision on effective frameworks and processes is of utmost importance for future results. Policies that might try to change outputs and outcomes only have little impacts if contexts or processes are ineffective and inefficient.

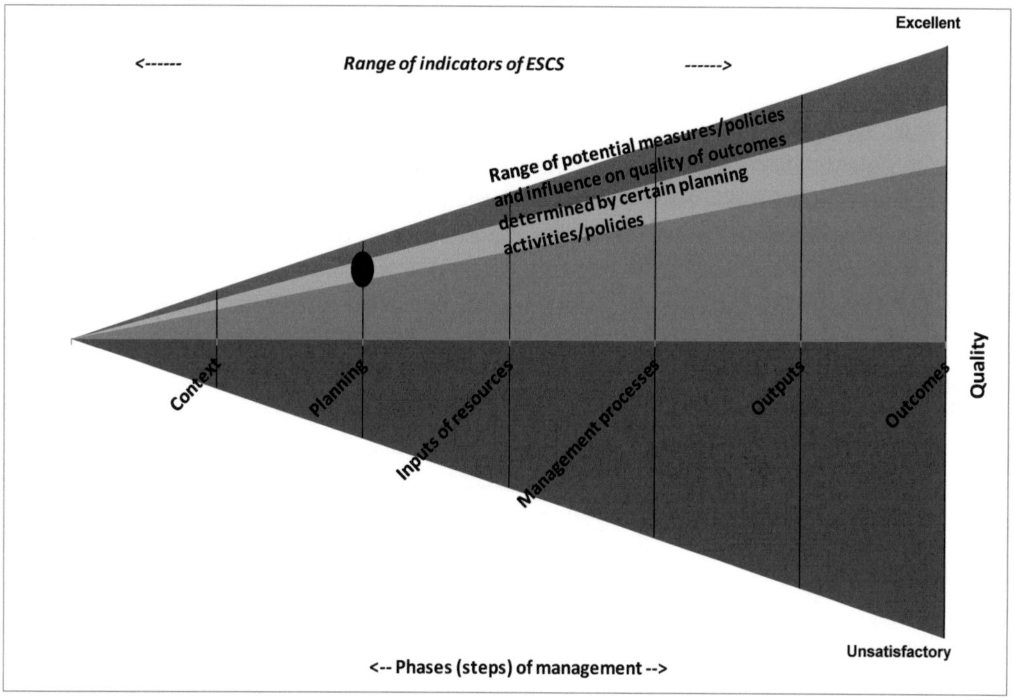

Source: Authors' draft based on Pfleger et al. (2009).

Figure 2. Long-term impacts of management decisions and room for improvements in a lock-in situation (management effectiveness triangle)

Regarding "paper parks" described above, there are examples of such parks that do not lack financial resource (e.g. international funding) but are designated without further activities. Evaluating effectiveness thus provides also a firm basis for discussing the financial inputs and their efficiency.

3. Linking PA effectiveness evaluation with an economic-ecological model of PA management, and governance and social learning

Evaluation of management effectiveness may be an important, even essential part in the management cycle of a protected area in particular with a focus on adaptive management strategies (Getzner et al., 2010). However, the current paper not only presents details of an evaluation of a national park (see section 4) but also extends the perspective of evaluation with respect to governance and social learning. Before discussing these aspects along the empirical case-study, we want to address evaluation from a systemic viewpoint linking evaluation with an economic-ecological model in a conceptual perspective.

Figure 3 presents such a conceptual model of the linkages between evaluation and monitoring by means of the European Site Consolidation Scorecard (ESCS) and the whole governance and financing system of a protected area. An ecological model of the ecosystem lies at the center of the conceptual framework describing the interdependencies between species (1), the habitats (2), and the abiotic features of the site. For simplicity, we assume that the species is an endemic animal species that may be observed in the habitat but is closely reliant on the quality and the size of the habitat since it cannot easily migrate to other habitats. In this ecological model (cf. Behrens et al., 2009), the species uses the habitat for food and reproduction, and underlies its own dynamics in terms of reproduction and decline which may seasonally change due to changing environments (provision of nutrients, water). This includes that the species is in competition with other species, and might also be a prey for natural enemies. We assume that protecting the animal species is crucial from a nature conservation perspective (e.g. endemic, or protected and listed in the species "red list"), and may easily be negatively affected by disturbing visitors. PA management may try to improve the reproduction of the species e.g. by temporal or permanent access bans to breeding grounds. The habitat itself has its own dynamics in terms of natural growth (including the abiotic elements) for which we assume that there exists a natural maximum size of the habitat in a certain area. The habitat may be impacted by visitors and their damages, and, again, by visitor management policies of the PA management.

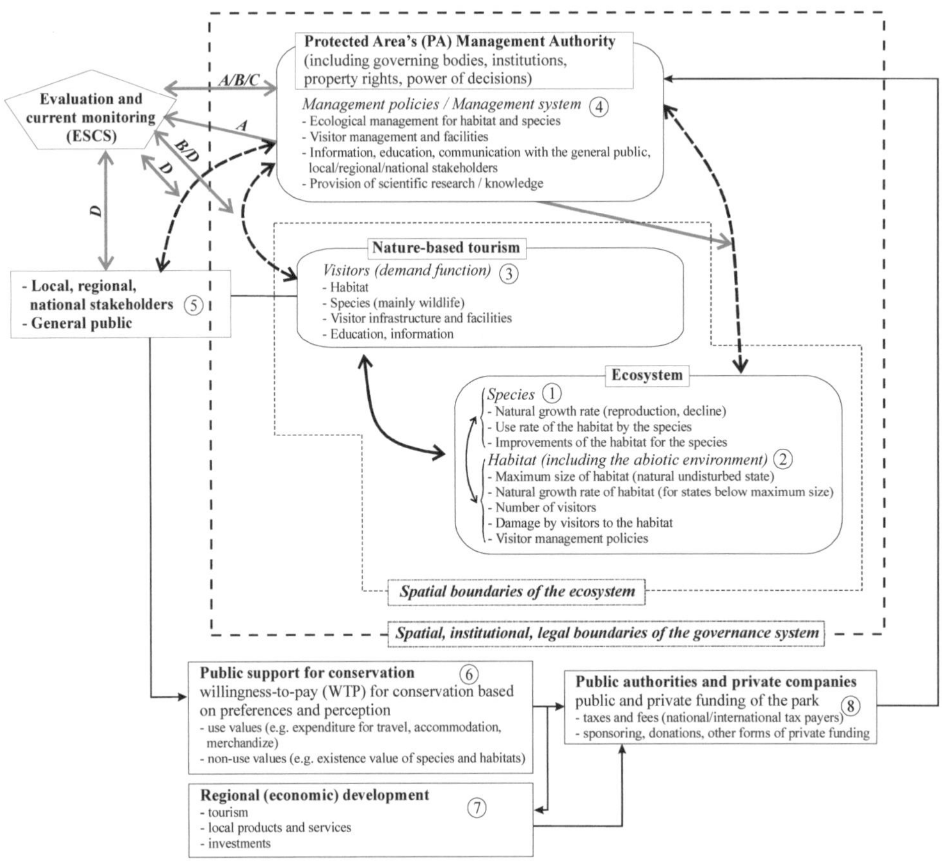

Source: Authors' own draft with reference to the ESCS, Bednar-Friedl et al. (2011), and Behrens et al. (2009).

Figure 3. Linking evaluation and monitoring according to the European Site Consolidation Scorecard (ESCS) with the protected area governance and financing system

The second part of the conceptual model may be labeled "economic" model since it refers to the economic use of the habitat in terms of demand for visits to the area (3). Visitors may be attracted to the area by a mixture of landscape (habitat) and species attributes (e.g. observation of wildlife). In addition, PA management can influence demand by providing or not providing (de-marketing) respective visitor infrastructure and facilities, and by education and information policies.

PA management, described by box (4), can steer (manage) the ecological, social and economic systems by a range of management policies such as ecological management for the habitat and species, and by aiming at the wide range of PA objectives such as visitor management and infrastructure provision, by information and education, and by production of scientific knowledge. Depending on the strength of the different objectives, PA management may have a certain objective function to be fulfilled (e.g. maximizing visitor benefits while sustaining a viable balance in the ecological system). In addition, PA management may also include the provision of non-use values (e.g. existence value of species/habitats) into its objective function.

The ecosystem, visitors, and the PA management – the latter including the whole range of property rights, relevant policies, decisions bodies and authorities – constitute the governance system of the protected area. Stakeholders (5) may partially be part of the governance system, for instance, by their inclusion in decision bodies of the protected area. However, we assume that the general public is outside the governance system.

The establishment and effective and efficient management of protected areas consumes resources, i.e., a range of costs have to be financed. On the one hand, the designation of a certain plot of land to be protected defines and changes property rights, and thus also alters the economic use of the area[3]. The costs associated with the designation of a protected area are therefore the opportunity costs (foregone benefits of using the area in an alternative way). In addition, "out-of-pocket" expenses such as personnel and equipment costs, investments for infrastructure, and other operating costs, have to be financed.

The efficiency and effectiveness of PA management, together with information provided, leads to a certain public support for the protected area in terms of willingness-to-pay for the financing the park (6). Willingness-to-pay may be measured by expenses of visitors (e.g. travel costs), and by the whole range of non-use values (e.g. existence values) of the general public. A positive attitude towards the protected area may also lead to a different regional economic development (7) which also may be a source of funding for PA management. Altogether, public authorities and private bodies (households, companies) pay for the establishment and operation of the protected area in terms of a range of contributions (e.g. taxes and fees, entry fees, sponsoring, donations) (8).

For understanding the implications of this conceptual model, it is important to consider that efficiency and effectiveness of PA management is paramount for securing funding. One the

[3] In the case of a national park according to IUCN's category II, all consumptive (extractive) uses of the area are prohibited.

one hand, the private sector is only willing to pay in terms of fees, travel costs, regional labeled products, etc., if PA management is effective and also sufficiently informs the general public. On the other hand, for securing public funds, national tax payers have to be convinced to provide sufficient financing.

Evaluation and current monitoring instruments thus may play a vital role in this conceptual model of managing and financing protected areas. The grey arrows and the letters (A to D) refer to the grouping of indicators of the European Site Consolidation Scorecard (ESCS) listed and described above (see Table 2). The different PA management policies are evaluated by different indicators, such as ecological management, visitor management, and education and information policies. However, as the ESCS is not only a tool for assessment based on data and statistics, the inclusion of several stakeholder groups is of specific importance. Stakeholder groups such as local decision-makers, experts, visitors, and PA business partners, in this context provide, first of all, information for attributing "correct" values to the indicators. Second, as stakeholders are involved by means of talks, discussions, and group work, the different perspectives become apparent. Third, the ESCS is also an information instrument by itself. By communicating the evaluation process and the results, the general public is informed about the use of public (and private) funds, and about the effectiveness of PA management.

In this conceptual model, evaluation thus becomes part of the governance system as an instrument of stakeholder involvement (participation), as well as an instrument of information. Furthermore, as the empirical case study will demonstrate, the ESCS process is a contribution to social learning and to "good governance". In general, accounting for stakeholder participation rests on the normative assumption that participatory frameworks lead to more efficient and effective park management. However, as governance systems including regulatory frameworks are different between regions and between parks, it is not straight forward to clarify the impacts of more or less participation. Hayes (2006) discusses this issue by presenting evidence both in favor and against participation of stakeholders by stressing that the "traditional" model of PA planning has recently regained more weight in effective park management. Schultz et al. (2011) argue in the same direction by cautiously describing how park management effectiveness might gain from participation of certain stakeholder groups (such as scientists, volunteers, and local inhabitants). The current chapter does not judge upon this question but rather shows that evaluation exercises might benefit from certain and well-defined instruments for stakeholder inclusion. As Raymond et al. (2010) point out, it is of specific importance in which frameworks local knowledge is collected and used for environmental management because evaluating management effectiveness might partially rely on tacit (local) knowledge for which stakeholder involvement is crucial.

Social learning in the context of (collaborative) resource management is defined as "learning that occurs when people engage one another, sharing diverse perspectives and experiences to develop a common framework of understanding and basis for joint action" (Schusler et al., 2003, 312). This general understanding of social learning emphasizes that – while the "learning unit" is the individual – learning is done through observation of or interaction with others in the social context, for instance, in a process of deliberation which "includes any process to communicate, raise and collectively consider issues, increase understanding, and arrive at substantive decisions" (Schusler et al., 2003, 312). Furthermore, social learning includes "learning by social aggregates, learning pertaining to social issues, and learning that results in recognizable social entities such as collective decision making procedures" (Maarleveld and Dangbégnon, 1999, 268). Even simple face-to-face communication has already proved to reduce resource consumption in experimental settings (Anderies et al., 2011), thus providing a fruitful starting point for social learning perspectives. Borowski et al. (2008) present a conceptual framework regarding the perspective of social learning in participatory resource management. They find that social learning may take place at many places in the governance and management processes (such as context, inputs, processes, outputs, outcomes) and provide indicators for the extent and effectiveness of social learning (see also Cheng et al., 2011). Consequently, social learning is supported by a range of different participatory instruments (e.g. information, stakeholder forums). As evaluation exercises provide several indicators for the efficiency and effectiveness of PA management, involvement of stakeholders in the evaluation process thus may also account for different forms of participation, from face-to-face discussions to workshops, and to comments and written critique in the evaluation report.

The ESCS methodology rests on a specific structure of indicators, and recommends a certain process how the information on each of the indicators is collected, discussed with stakeholders, condensed and assessed, and summarized to a certain score value. Insofar, evaluation does not seem to provide a major contribution to social learning allowing for extensive deliberative processes. While the ESCS may not be able to make up for missing deliberation and participation in the governance structure of the PA management – the lack of such participatory structures should become apparent by the evaluation itself – the evaluation process might also constitute an element of deliberation due to the inclusive nature of the indicator assessment, with the potential aim to encourage "social partnerships" among stakeholders and the park management (Benn, 2010). As such, an evaluation of management effectiveness in parks (e.g. by means of the ESCS) may also form one part of a governance structure of a protected area. Governance in the context of protected areas is usually defined as the "interactions among structures, processes and traditions that determine how power is exercised, how decisions are taken, and how citizens or other stakeholders have their say" (Graham et al., 2003, 2f.). Decision-making rules, including responsibility and accountability, are clearly part of the governance system of protected areas. While the leeway within an evaluation exercise is certainly limited since evaluation has to be based on a certain methodology and structure which cannot be changed during the evaluation process without loss of credibility and focus, the applied evaluation instruments allow for different extents of participation and deliberation.

Whether evaluation instruments in principle contribute to "good" governance is an additional question. The *positive* question of research concerning the role of evaluation instruments in the governance system would be generally answered by assessing how the governance framework is set up (system of rules, regulatory structure; cf. Mayntz, 2005; Ostrom, 1990), what the roles the different stakeholders have, and how the system works. The *normative* question refers to the quality of governance and potential recommendations for improving the frameworks. As Getzner et al. (2012) write, "good" governance might "crucially influence whether the protected area can achieve its objectives, is able to fairly share benefits and costs, and seeks and gains sufficient support from local communities and stakeholders" (cf. Lockwood, 2010).

Building up social capital by learning due to appropriate stakeholder involvement is nevertheless one important ingredient in participatory resource management (Enengel et al., 2011).

4. Evaluating management effectiveness at Gesäuse national park (Austria)

4.1. Choice of methodology and case study area

Based on a feasibility studie (Jungmeier & Velik, 1999) and an intensive regional public debate the Gesäuse national park was established in the Austrian federal state of Styria in 2002 as a national park conforming to IUCN's category II of protected areas. Only in 2003, the national park was officially acknowledged by IUCN to conform to category II. The landscape is alpine with deep gorges formed by the Enns river and its contributing streams with small patches of high-diversity wetlands adjacent to the river. The area of the national park amounts to around 11,000 hectares of which 86% are core zones (the rest consists of buffer zones and sustainable pastures). The park stretches from around 500 to over 2,300 meters above sea level, with about 63% of the area consisting of forests, 30% high alpine areas, and the rest water bodies and pastures. The land is almost entirely owned by the public (federal state of Styria). Four rural municipalities form the core of the "Gesäuse national park region". The national park administration is also responsible for managing the area according to the European Union's Natura 2000 network.

The park was established in 2002, based on an agreement between the federal state of Styria and the Republic of Austria (Ministry of the Environment) sharing the management and financial responsibilities. While this agreement – codified in a regional law – comprehensively deals with all aspects including IUCN management criteria, it also provides the legal basis for regularly evaluating management effectiveness of the park. According to the law, comprehensive evaluations have to take place once every five years with the aim to assess management effectiveness in all dimensions of the park's organizational and managerial fields, such as the organization, financial and activities planning, ecological management, visitor policies, and scientific research.

The current evaluation project was commissioned in 2007 by the national park authority to an interdisciplinary group consisting of M. Jungmeier (ecologist and nature conservation planner for over 20 years), B. Pfleger (environmental engineer and conservation planner/manager), W. Scherzinger (professor of ecology and nature conservation), and M.

Getzner (ecological economist). The framework of evaluation was set up by the European Site Consolidation Scorecard (ESCS) with the explicit recommendation to consult all relevant stakeholders during the evaluation process, and the request by the park management to present a very detailed assessment of the park.

The ESCS evaluates the effectiveness of management in a broad perspective, addressing all relevant issues of a park's management, and measuring the success or failure by means of detailed scores for each indicator. A brief description of the methodology can be found in Pfleger, 2010. Table 2 above presents an overview of the indicators of this evaluation tool.

Each indicator is evaluated according to a pre-defined scale; for instance, the indicators measure whether the financial basis of the park is sustainable and secured for the future, and whether effective management plans are developed and implemented. Five different levels of implementation and effectiveness are described, explained, and can be selected. In addition, management tools, references and best practice examples for each indicator are included in the description of the ESCS framework and instrument. Finally, there is a comprehensive documentation section for recording information, and, among others, for verifying the results and helping to implement necessary measures and policies.

The measurement of indicators along a scorecard is probably not the only significant contribution of the ESCS evaluation tool. While the assigning of certain scores to the indicators was done by the research team, the concrete scores for the indicators are not the most relevant element of the evaluation process. The first step of the evaluation was to collect a comprehensive informational basis for assessing management effectiveness which was done by contacting all relevant stakeholders; at the beginning, information was, of course, collected with the national park authorities including the management, and the local nature conservation authorities. The informational basis was, as objectively as possible, described and condensed for each indicator. The second step was to collect stakeholders' opinions regarding the current state of each indicator. For some indicators, different stakeholder groups presented their assessment of the current situation which was documented in the ESCS report, too. The third working step included an assessment from the viewpoint of the expert (evaluation) team, and the final (fourth) section of each indicator assessment included a comprehensive list of recommendations for further improving management effectiveness.

It is of high importance to note that the evaluation of management effectiveness by means of evaluation tools, especially in the case of the ESCS, is concentrated towards the long-term development of the park. In the case of the Gesäuse national park, the evaluation of the long-term sustainability of current policies, including the frameworks mentioned above, is particularly significant. For instance, current policies in transforming the composition of tree species in the diverse forests of the park can only be assessed in the far future due to the slow development of forests towards a natural composition and ecological balance. As national parks secure and partially (re-) introduce natural dynamics to the ecosystems, these processes are inherently stretching far into the future. A basic assessment such as the ESCS therefore cannot, and does not want to, assess management effectiveness in the sense of measuring the current benefits of policies (outputs and outcomes), as this would often require long-term monitoring and research. Instead, the ESCS has the aim to review current policies with the perspective of the sustainability and future achievements of PA authorities. Based on the assumption, that changes in the first phases of the management cycle (see Figure 2) have far more significant impacts on the effectiveness of a protected area than improvements in the latter ones, the ESCS focuses on the first four phases (context, planning, inputs, processes), rather than on concrete outcomes. The last phases are indirectly addressed by the respective indicators for comprehensive monitoring which measures the objectives of the park.

As mentioned above, a particularly strong focus of the ESCS was the involvement of local and regional stakeholders. At the time of evaluating the park, park policies were subject to many emotionalized debates in the region. The need to identify and integrate critical perceptions into the evaluation was obvious, but a clear distinction between opinions and facts was considerably difficult. To secure the credibility of the process and the whole evaluation efforts, the discussion with stakeholders was structured according to the "Regional Timeline Method" (Jungmeier et al., 2010). Based on individual events and incidents, stakeholders created a joint emotional timeline for particular periods from the inauguration of the park until the evaluation exercise. This method thus complemented the evaluation process by providing "screenshots" of stakeholders' views.

4.2. Results of the evaluation

As mentioned before, evaluating the Gesäuse national park is included in the governance framework of the park which is itself codified in the regional national park law. The evaluation was carried out based on international standards with reference to the existing evaluation frameworks while applying a newly developed, participatory evaluation methodology called "European Site Consolidation Scorecard" (ESCS). It is safe to say that this evaluation exercise has been by far the most in-depth evaluation for Austrian parks to date. The analysis was based on all important documents, personal talks, several workshops in the region, and an assessment of the regional economic impacts. The evaluation report summed up the results in the following way. The Gesäuse national park management can be described as "exceptionally positive. The foundations that are laid down, the developed structures in the park, processes and activities certainly compare favorably in the international context. Strategic decisions of the park management are well argued, the documentation is comprehensive, and the overall direction of the park is sustainable." (Getzner et al., 2008; translation from German).

In particular, the following fields and projects of the national park management are exceptional:

- Innovative and high-quality, partially unique offers for visitors regarding education and information (visitor center, exhibitions, events);
- Comprehensive and systematic public relations work within and outside the region;

- Highly qualified, motivated and effective team members and team work;
- Comprehensive basic ecological inventory, excellent scientific research and planning work;
- First already verifiable impacts on the regional economy due to increases in visitor numbers.

On the other hand, the evaluation highlighted some aspects that might need further concentration of resources:

- Reduction of the barrier effect of roads, rail tracks, forestry and provisioning paths;
- Assurance of financial sustainability by valorization of public budgets based on the consumer price index;
- Further development of the legal frameworks to stringently apply IUCN criteria;
- Improvement of participatory processes for stakeholders and vested interests inside and outside the region;
- Development of a comprehensive and inclusive management plan with the park's vision, objectives, and implementation strategies;
- Improvement of collaboration between the land owner (Styrian Forests, a publicly owned company) and the national park authority;
- Improvement of the boundaries of the national park in the light of ecological dynamics.

All in all, the existing evaluation of the Gesäuse national park has certified that the park's management has decided upon sustainable strategies for improving and securing the natural dynamics in the ecosystems while providing excellent infrastructure for visitors and information for stakeholders and the general public.

5. Discussion, summary and conclusions

During the life-cycle of a protected area (PA), the evaluation of management effectiveness of the PA administrating and managing authorities including the whole PA context becomes increasingly important, both for securing and improving the conservation of biodiversity, but also for the acceptance of stakeholders and funding bodies. Currently, there are numerous approaches to evaluating management effectiveness of parks; many international institutions have drafted and implemented such evaluation instruments.

The current paper has focused on a specific evaluation instrument that emphasizes the process of evaluation from the perspective of a range of important stakeholders. The "European Site Consolidation Scorecard" (ESCS) is an evaluating instrument with four large groups of indicators dealing with strategic planning, conservation activities and results, long-term financing, and site constituency. The instrument thus has become part of the governance system of the park contributing to social learning both of the PA staff as well as local stakeholders in manifold dimensions.

At the beginning of any evaluation exercise, PA management and policy makers have to be convinced of the usefulness of the evaluation to increase the probability that recommendations will indeed be implemented (Pfleger, 2008). The implementation of the evaluation results is crucial since the evaluation might lead to higher costs than benefits since the frustration of those involved in the evaluation process might be significant (Hockings et al., 2006). However, the lack of implementing evaluation results is, at the moment, one of the biggest problems in assessing protected area management effectiveness. Evaluations are carried out, but recommendations are not implemented in daily management (Steindlegger, 2007).

In general, external evaluation is important because it addresses "uncomfortable" topics such as the organizational structure, enhances credibility of park management, and forms the basis for policy changes. However, it is useless without sufficient discussion and participation of the PA management because the administration of the protected area has the responsibility to implement the recommendations. Without understanding evaluation results in a respective social learning environment, PA management will not follow recommendations and adapt management policies.

Integrating stakeholders' opinions and viewpoints is necessary especially for collecting information about PA policies and impacts. In addition, it is of high importance to present stakeholders with the opportunity to express their opinions with the aim to making PA management more effective, and to increase the acceptance and support of the evaluation results. Suitable instruments may be workshops, and personal talks and interviews as an addition to the existing governance framework. However, once stakeholders are involved, their opinion has to be accounted for in a transparent way. In addition, evaluation results have to be made available for stakeholders (as well as the general public) to receive a sufficient acceptance of evaluation results. Publicizing evaluation results has to be done in a sensitive way because presenting bare scorecard figures to the general public may lead to misinterpretations.

As the European Site Consolidation Scorecard (ESCS) focuses on the first four phases of the management cycle and not on outputs and outcomes, the evaluation method cannot assess the efficiency of PA management in terms of relating outcomes to the costs of the park. The ESCS thus has to be complemented by long-term monitoring methods (however, the ESCS indirectly addresses long-term developments and outcomes by including an indicator assessing the PA's monitoring plan). During the detailed ESCS evaluation process in the Gesäuse national park some crucial elements of effective PA management such as the missing management plan or insufficient stakeholder involvement, were soon obvious. As a methodological conclusion of our evaluation study, it is not necessary to employ in-depth evaluation tools for discovering the most obvious weaknesses of PA management which might be evident by using a simple scorecard with some stakeholder involvement. Thus, a first quick evaluation might exhibit crucial issues with relative low efforts if the evaluation concentrates on the first four management phases. If results should be more comprehensive, and provide more detailed recommendations, costly evaluation and long-term monitoring tools are necessary.

Whether protected area management effectiveness evaluations really contribute to good governance and social learning depends less on the methodology chosen (whether e.g. it is a rapid assessment or an in-depth evaluation). Instead, the evaluation process itself is of much more importance. A "secret" self-assessment done by the protected area management body will hardly be useful whereas an integrated approach that also includes external experts as well as relevant stakeholders, and widely publishes the results, brings far more benefits than just management recommendations.

The debate on appropriate indicators, evaluation tools, scorecards and checklists, has intensified in the last decade but remains at a rather technical level. However, any learning and progress in and about protected areas management is clearly connected to the intensity of the public debate provoked by the evaluation process. The evaluation of the Gesäuse national park as part of the park's governance had mainly three results. First, a written evaluation report including the scorecard and the recommended policies; second, a broad range of personal viewpoints and findings of over 50 staff members and stakeholders directly involved in the evaluation exercise in the sense of social learning experiences; third, a broader understanding of the PA staff for the principles of adaptive management that may quickly respond to changing needs and requirements. Summing up, the authors of the current study consider the first result – while the only "tangible" output of the evaluation – to be the least important one.

Author details

Michael Getzner
Center of Public Finance and Infrastructure Policy,
Vienna University of Technology, Vienna, Austria

Michael Jungmeier
ECO Institute of Ecology, Klagenfurt, Austria

Bernd Pfleger
Experience Wilderness, Enns, Austria

Acknowledgement

We are thankful to the editor and the reviewers for their comments and suggestions. In addition, we would like to thank W. Scherzinger, W. Franek, and the Gesäuse national park staff, for their collaboration and support during the evaluation of the national park. All errors are, of course, the responsibility of the authors.

6. References

Anderies, J. M., Janssen, M. A., Bousquet, F., Cardenas, J.-C., Castillo, D., Lopez, M.-C., Tobias, R., Vollan, B., Wutich, A. (2011). The challenge of understanding decisions in experimental studies of common pool resource governance. Ecological Economics 70, 1571-1579.

Bednar-Friedl, B., Behrens, D., Getzner, M. (2011). Socioeconomics of conservation in the Alps. In: Schmidt, J. G. (ed.), The Alpine Environment: Geology, Ecology and Conservation. NOVA Science Publishers, Hauppauge (NY), 135-152.

Behrens, D., Friedl, B., Getzner, M. (2009). Sustainable management of an alpine national park: handling the two-edged effect of tourism. Central European Journal of Operations Research 17 (2), 233-253.

Benn, S. (2010). Social partnerships for governance and learning towards sustainability. ARIES Working Paper 1/2010, The Australian Research Institute for Environment and Sustainability (ARIES), Department of Environment & Geography, Macquarie University, North Ryde (NSW).

Bonham, C. A, Sacayon, E., Tzi, E. (2008). Protecting imperiled "paper parks": potential lessons from the Sierra Chinajá, Guatemala. Biological Conservation 17, 1581-1593.

Borowski, I., le Bourhis, J.-P., Pahl-Wostl, C., Barraqué, B. (2008). Spatial misfit in participatory river basin management: Effects on social learning, a comparative analysis of German and French case studies. Ecology and Society 13, paper 7 (www.ecologyandsociety.org, 17 April 2012).

Brandon, K., Redford, K. H., Sanderson, S. E. (eds.) (1998). Parks in peril: people, politics, and protected areas. Island Press, Washington (D.C.).

Bruner, A. G., Gullison, R. E., Rice, R. E., da Fonseca, G. A. B. (2001). Effectiveness of parks in protecting tropical biodiversity. Science 291, 125-128.

Chape, S., Harrison, J., Spalding, M., Lysenko, I. (2005). Measuring the extent and effectiveness of protected areas as an indicator for meeting global biodiversity targets. Philosophical Transactions of the Royal Society B: Biological Sciences 360, 443-455.

Cheng, A. S., Danks, C., Allred, R. S. (2011). The role of social and policy learning in changing forest governance: An examination of community-based forestry initiatives in the U.S. Forest Policy and Economics 13, 89-96.

Diller, C. (2008). Evaluierungen und Regional Governance: Funktionen der Evaluierung von und in regionalen Steuerungsstrukturen. Zeitschrift für Evaluierung 8, 273-301.

Enengel, B., Penker, M., Muhar, A., Williams, R. (2011). Benefits, efforts and risks of participants in landscape co-management: An analytical framework and results from two case studies in Austria. Journal of Environmental Management 92, 1256-1267.

European Council (1992). Council Directive 92/43/EEC of 21 May 1992 on the conservation of natural habitats and of wild fauna and flora. EU Official Journal L 206, 22 July 1992, 7.

Gantioler, S., Rayment, M., Bassi, S., Kettunen, M., McConville, A., Landgrebe, R., Gerdes, H., ten Brink, P. (2010). Costs and socio-economic benefits associated with the Natura 2000 network. Final report to the European Commission, DG Environment on Contract ENV.B.2/SER/2008/0038. Institute for European Environmental Policy / GHK / Ecologic, Brussels.

Getzner, M., Jungmeier, M., Pfleger, B., Scherzinger W. (2008). Evaluierung Nationalpark Gesäuse. Report of E.C.O. Institute of Ecology, Klagenfurt (Austria).

Getzner M., Jungmeier, M., Lange S. (2010). People, Parks and Money. Stakeholder Participation and Regional Development – a Manual for Protected Areas. Heyn Verlag, Klagenfurt.

Getzner, M., Jungmeier, M. (2012). The contribution of protected areas to sustainability. Journal of Sustainable Society (forthcoming).

Getzner, M., Lange Vik, M., Brendehaug, E., Lane, B. (2012). Governance and management strategies in national parks: Implications for sustainable regional development. International Journal of Sustainable Society (forthcoming).

Graham, J., Amos, B., Plumtre, T. (2003). Governance principles for protected areas in the 21st century. Paper presented at the 5th World Parks Congress, Durham, South Africa.

Hayes, T. (2006). Parks, people, and forest protection: An institutional assessment of the effectiveness of protected areas. World Development 34, 2064-2075.

Hockings, M., Stolton, S., Leverington, F., Dudley, N., Courrau, J. (2006). Evaluating effectiveness: a framework for assessing management effectiveness of protected areas. 2nd ed., IUCN, Gland (Switzerland)/Cambridge (UK).

Jungmeier, M., Kirchmeir, H., Kühmaier, M., Velik, I., Zollner, D. (2005). IPAM-Toolbox. Transnational Results (Expert System, Toolbox and Best Practice). Report for the Government of Carinthia (Austria), E.C.O. Institute of Ecology, Klagenfurt.

Jungmeier, M., Paul-Horn, I., Zollner, D., Borsdorf, F., Lange, S., Reutz-Hornsteiner, B., Grasenick, K., Rossmann, D., Moser, R., Diry, Ch. (2010). Partizipationsprozesse in Biosphärenparks – Interventionstheorie, Strategieanalyse und Prozessethik am Beispiel vom Biosphärenpark Wienerwald, Großes Walsertal und Nationalpark Nockberge (Band I: Zentrale Ergebnisse). Austrian MAB National Committee and Austrian Academy of Sciences, Vienna.

Joppa, L. N., Loarie, S. R., Pimm, S. L. (2008). On the protection of "protected areas". Proceedings of the National Academy of Sciences (PNAS) 105, 6673-6678.

Kirchmeir, H., Zak, D., Getzner, M. (2012). Gap analysis: Individuation of gaps in the management and implementation of Natura 2000 sites. Report for the SEE Be-Natur project, Vienna University of Technology.

Leverington, F., Costa, K. L., Courrau, J., Pavese, H., Nolte, C., Marr, M., Coad, L., Burgess, N., Bomhard, B., Hockings, M. (2010a). Management effectiveness evaluation in protected areas – a global study. 2nd ed., School of Geography, Planning and Environmental Management, The University of Queensland, Australia.

Leverington, F., Costa, K. L., Pavese, H., Lisle, A., Hockings, M. (2010b). A global analysis of protected area management effectiveness. Environmental Management 46, 685-698.

Lockwood, M. (2010). Good governance for terrestrial protected areas: A framework, principles and performance outcomes. Journal of Environmental Management 91, 754-766.

Maarleveld, M., Dangbégnon, C. (1999). Managing natural resources: A social learning perspective. Agriculture and Human Values 16, 267-280.

Mayntz, R. (2005). Governance Theory als fortentwickelte Steuerungstheorie? In: Schuppert, G. F. (ed.), Governance-Forschung: Vergewisserung über Stand und Entwicklungslinien. Nomos, Baden-Baden, 11-20.

Ostrom, E. (1990). Governing the commons: the evolution of institutions for collective action. Cambridge, Cambridge University Press.

Pfleger, B. (2007a). Evaluation of the management effectiveness of Central European protected areas – A critical revision of the Parks in Peril Site Consolidation Scorecard. Master thesis (Management of Protected Areas Programme), University of Klagenfurt.

Pfleger, B. (2007b). European Site Consolidation Scorecard (ESCS) – Measuring the management effectiveness of European protected areas. Klagenfurt University / E.C.O. Institute of Ecology, Klagenfurt (Austria), mpa.e-c-o.at (16 April 2012).

Pfleger, B. (2008). Management effectiveness assessment in Austria – site consolidation scorecard. In: Stolton, S. (ed.), Assessment of management effectiveness in European protected areas. Proceedings of a Seminar organized by BfN and EUROPARC Federation on the Island of Vilm, Germany, April 2008, 46-48.

Pfleger, B., Jungmeier, M., Hasler, V., Zacherl-Draxler, V. (2009). Leitfaden zur Evaluierung des Nationalparkmanagements in Österreich - Report of E.C.O. Institute of Ecology, Klagenfurt (Austria)

Pfleger, B. (2010). European Site Consolidation Scorecard, Austria. In: Leverington, F., et al. (eds.), Protected Area Management Effectiveness Assessments in Europe – Supplementary Report: Overview of European methodologies, 68-71.

Pomeroy, R. S., Watson, L. M., Parks, J. E., Cid, G. A. (2005). How is your MPA doing? A methodology for assessing management effectiveness of marine protected areas. Ocean and Coastal Management 48, 485-502.

Powell, R. B., Vagias, W. M. (2010). The benefits of stakeholder involvement in the development of social science research. Park Science 27, 46-49.

Raymond, C. M., Fazey, I., Reed, M. S., Stringer, L. C., Robinson, G. M., Evely, A. C. (2010). Integrating local and scientific knowledge for environmental management. Journal of Environmental Management 91, 1766-1777.

Schultz, L., Duit, A., Folke, C. (2011). Participation, adaptive co-management, and management performance in the World Network of Biosphere Reserves. World Development 39, 662-671.

Schusler, T. M., Decker, D. J., Pfeffer, M. J. (2003). Social learning for collaborative natural resource management. Society and Natural Resources 15, 309-326.

Secretariat of the CBD (2005). Handbook of the Convention on Biological Diversity including its Cartagena Protocol on Biosafety. 3rd edition, Secretariat of the Convention on Biological Diversity (CBD), Montreal.

Steindlegger, G. (2007). WWF International – Member of the WWF Global Protected Area Team. Personal communication.

Stockmann, R. (2007). Evaluation in der Gesellschaft: Entwicklung, Stand und Perspektiven. Zeitschrift für Evaluation 6, 195-222.

Stoll-Kleemann, S. (2010). Evaluation of management effectiveness in protected areas: methodologies and results. Basic and Applied Ecology 11, 377-382.

The Nature Conservancy (2004). Measuring success: The Parks in Peril Site Consolidation Scorecard manual. www.parksinperil.org (4 November 2006).

WDPA (2012). Protected Areas Management Effectiveness Information Module. WDPA (World Database on Protected Areas), www.wdpa.org (16 April 2012).

Chapter 8

BE-NATUR: Transnational Management of Natura 2000 Sites

Renate Mayer, Claudia Plank, Bettina Plank,
Andreas Bohner, Veronica Sărățeanu, Ionel Samfira,
Alexandru Moisuc, Hanns Kirchmeir, Tobias Köstl,
Denise Zak, Zoltán Árgay, Henrietta Dósa, Attila Gazda,
Bertalan Balczó, Ditta Greguss, Botond Bakó, András Schmidt,
Péter Szinai, Imre Petróczi, Róbert Benedek Sallai, Zsófia Fábián,
Daniel Kreiner, Petra Sterl, Massimiliano Costa, Radojica Gavrilovic,
Danka Randjic, Viorica Bîscă, Georgeta Ivanov and Fănica Başcău

Additional information is available at the end of the chapter

http://dx.doi.org/10.5772/50653

1. Introduction

The EU-wide network of Natura 2000 sites serves the conservation as well as the protection of endangered species and habitats. One must differentiate between Special Areas of Conservation (SAC), which are protected by the Habitats Directive 92/43/EEC, and Special Protected Areas (SPA), protected by the Birds Directive 2009/147/EC.

All nature sites have an obligation to report to the EU. The development of the Natura 2000 sites (provisions, effects) must be documented in three- or six-year spans. Member states must present conservation plans for all sites of the Natura 2000 network and conduct monitoring. This is to provide information about the preservation of the conditions of the protected species and habitats. How does it affect the protected property? Did the provisions get implemented? Did they have the desired success? Are additional provisions necessary? A habitat is considered adequately covered if up to 60 per cent of its total area is contained within recommended sites. If less than 20 per cent of a habitat's total area is contained in the sites, it is insufficiently represented. The countries are obliged to implement the Natura 2000 standards, or else they face the penalty of high fines to the EU.

According to the first reporting period of inquiry from 2001 to 2006, only 17 per cent of the already reported Natura 2000 areas in the entire EU were in a favourable state of preservation. The rest were either unfavourable/bad, unfavourable/inadequate, or unknown. Assessments of the state of conservation were not carried out on 18 per cent of the areas. The state of conservation of habitat types associated with agriculture have was worse, with only seven per cent of assessments being favourable, compared to 21 per cent for non-agricultural habitats, in [1].

The Interreg South-East-Europe project BeNatur (Better Management of Natura 2000 Sites) aims at improving the management and organisation of those Natura 2000 sites, with particular focus on wetland areas. In view of actually implementing the EU legal framework in this regard, measures like the definition and implementation of a Joint Transnational Strategy for better management and improvement of the Natura 2000 network, the definition of Joint Transnational Action Plans for the conservation of the habitats and species common to partner areas, practical pilot projects for implementation of the Joint Transnational Strategy and communication activities to build awareness on environment protection are implemented in the partner countries. As basis for the project a gap analysis, which is an approach for the assessment of gaps of existing protected area networks, was made, see [2].

2. Biodiversity in the grassland ecosystem

The term biodiversity is perceived differently among professional groups, such as specialists in taxonomy, economists, agronomists, or sociologists. Moreover, biologists tend to define biodiversity as the diversity of all living beings. Farmers are interested in exploiting the potential arising from the diversity of soils, regions, etc. For industry, biodiversity represents a reservoir of genes of use for biotechnologies or exploitable resources such as timber, fish, etc. The general public largely associates biodiversity to exceptional landscapes or endangered species.

2.1. Defining the concept and levels of biodiversity

In the 1960s, the understanding of ecosystems, as known today, began to rely upon the functioning of natural systems in relation to different species of animals and plants. In the early 70s, specialists started the debate whether certain species may be considered useful or harmful. The term biodiversity (ecological diversity) denotes the variety of life forms which inhabit the biosphere. This is measured by the total number of species (plants, animals, fungi, microorganisms) which represent all terrestrial and aquatic ecosystems found on the planet. According to [3], biodiversity can be considered on five levels: ecosystem diversity; diversity of biological systems in time; assemblies of species diversity; specific diversity; genetic diversity (intraspecific diversity).

Number of species from a particular biological community is expressed as species richness or alpha diversity. This can be used to compare the number of species in different geographic areas or biological communities. Beta diversity refers to the gradient along which the structure of species changes in relation to a geographic factor. Gamma diversity is characteristic of regions and includes the biodiversity of ecosystems which make up the characterized region.

2.2. Intraspecific (genetic) biodiversity

Field studies of various species confirm the existence of differences between conspecific individuals. These differences often intensify along with the extension of the area occupied by that particular species, see [4].

A population is characterized by a particular genetic structure expressed in gene frequency. The difference between populations of the same species is determined precisely by differences in gene frequency. The sources of genetic variability are mutations, genetic recombination, and gene flow, all acting in relation to the demands of selection, which finalizes the nature of the variations. The diversity of spontaneous or cultivated species' populations is studied by several methods, among which the most commonly used are the following: morphological descriptive methods; methods based on phonological and physiological characters; cytogenetic methods; methods based on the use of molecular markers.

2.3. Species biodiversity

It is estimated that biological communities developed over millions of years are breaking down due to anthropic influence. A large number of species have declined rapidly, some of them being on the verge of extinction. The causes for this situation are multiple, ranging from excessive hunting and habitat destruction to the attack of predators or competitors introduced by humans.

Numerous studies on natural and artificial grasslands confirm the fact that loss of species generates a decline of productivity, the community being less able to respond to environmental variations (e.g. drought). When species are lost, biological communities display a substantially reduced resilience to changes caused by human activities.

2.4. Ecosystem biodiversity

Ecosystem biodiversity refers to the complexity of structural elements (species) and functional elements which characterize an ecosystem on the level of spatial and temporal units. The most prominent manifestation of ecosystem biodiversity is conveyed by species richness (species composition). The species richness of an ecosystem is a result of the interactions between species and the biotope, as well as an effect of interspecific relations.

The biotope displays a selective action upon the species from the early stages of its colonization, eliminating species whose adaptations do not correspond to the variation regime of abiotic factors. Thus, biotope acts as a filter (biotope filter).

If biomass production is low, there is intense competition for nutrients and only slow-growing species are able to survive. If production is high, only a few fast-growing species resist in the competition for light, nutrients, etc.

2.5. Methods for quantifying biodiversity

Biodiversity can be classified according to different criteria, e.g. from a taxonomic or ecological point of view. At a taxonomic level, biodiversity is regarded as the species richness of a given area at the moment of the examination.

From an ecological point of view, in [5], biodiversity is characterised on different spatial scales, as follows: alpha diversity, beta diversity and gamma diversity represents the quantification of biodiversity in all ecosystems from a region.

Numerous diversity indices are described in the specialty literature. One category of diversity indices takes into account the total number of species (S) and the total number of individuals (N) present in the sampling surface.

2.5.1. Alpha biodiversity

Alpha biodiversity refers to the biodiversity of a given surface area or of an ecosystem and it is generally expressed as the number of species from that ecosystem, see [5].

2.5.2. Species richness

Species richness is an insufficient parameter to determine ecosystem biodiversity (Barbault, 1992). Two ecosystems may display the same degree of biodiversity if the number of species they contain is equal, although some species may be represented by fewer individuals. It is therefore necessary to determine quantitative indices for species assessment, such as abundance and frequency.

2.5.3. The Shannon indices

The Shannon indices characterize the plant community from a taxonomic point of view, respectively in terms of specific diversity. The two indices are entropy (H') and relative abundance ($E_{H'}$), also called equitability index. The Shannon index (H') employs specific data to measure biodiversity, as it characterizes the entropy of the analyzed sample, considering the species as a symbol and the relative size of their populations as probability.

The Shannon equitability index ($E_{H'}$) represents the relative abundance of the studied sample on a scale from 0 to 1, where values close to 0 indicate a low regularity of species presence and 1 stands for the maximum species' regularity of occurrence. This index is calculated by the formula: $E_{H'}=H'/H_{max}=H'/\ln S$

2.5.4. Simpson's indices

Simpson's index takes into account not only the number of species but also the proportion of each species. Much like the indices presented in the previous section, Simpson's diversity indices represent a means of assessment which takes into account species richness and the species' regularity of occurrence in the sample. Simpson's index (D) measures the probability that two individuals randomly taken from a sample belong to the same species. $D = \sum (n_i / N)^2 = \sum_{i=1}^{S} p_i^2$ where: n_i = the total number of individuals of the species i;N = the total number of individuals of all species in the sample;$p_i = n_i / N$.

2.5.5. Beta biodiversity

Gamma biodiversity takes into consideration changes in species diversity between different ecosystems. Generally, the total number of species in each analyzed ecosystem is determined and they are compared by assessing the total number of different species [5]. Ecological similarity indices (floristic similarity indices) are employed in order to compare two samples of grassland floristic relevés. These fall into two categories, namely: qualitative and quantitative similarity indices.

2.5.6. The Jaccard index

The Jaccard Index (P_J) also called the Jaccard similarity coefficient was introduced by Paul Jaccard (1902, 1928) to statistically compare the similarity and diversity between sample sets. This coefficient allows comparison between station fields based on the following mathematical formula and it can have values ranging from 0 to 100: $P_J = [c/(a + b - c)] \times 100$ (original formula) where: a – the number of species from a list (i.e. relevé 1); b – the number of species from a list (i.e. relevé 2); c – the common number of species (i.e. in the two compared relevés).

2.5.7. The Sørensen index

The Sørensen index (Q_S), also known as Sørensen's similarity coefficient, was developed by botanist Thorvald Sørensen in 1948. Like the Jaccard index, this similarity index takes into account the presence or absence of data. The following formula is employed to determine this index: $Q_S = 2c/(a + b)$ where: a – the number of species from a list (i.e. relevé 1); b – the number of species from a list (i.e. relevé 2); c – the common number of species (i.e. in the two compared relevés).

2.5.8. The Hamming distance

The notion of Hamming distance (H) was proposed by Daget & al. (2003), cited by Le Floc'h (2007), for comparing floristic relevés by utilizing the following formula: $H = 1 - P_J$ where: P_J – the Jaccard index.

2.5.9. The Renkonen index

The Renkonen index (P), also known as the percent similarity, was developed around 1938. It employs the abundance of species from the compared relevés in applications on grassland vegetation and the obtained result is expressed as a percentage. The Renkonen index is calculated by the following formula: P = Σmin(p_{1i}, p_{2i}) where: p_{1i} – frequency of species i in relevé 1; p_{2i} – frequency of species i in relevé 2.

2.5.10. Gamma biodiversity

In [6] Gamma biodiversity represents the general diversity of all ecosystems which make up a region, respectively the diversity of species on a global scale [6]. According to [7], $\alpha \times \beta = \gamma$, a relation which must be valid for all the indices utilized in the calculation, according to [8]. Thus, gamma diversity must be determined entirely by alpha and beta diversity. Jost does not specify in what manner alpha and beta determine gamma, but mentions that alpha diversity can never be higher than gamma diversity.

2.5.11. The Mean Species Abundance Index (MSA)

The Mean Species Abundance Index (MSA) is currently used to determine gamma diversity and biodiversity on a regional scale. It was developed under the coordination of the Convention on Biological Diversity (CBD) and it provides information on presently extant biodiversity compared to that of previous periods.

The MSA index is used to calculate the estimated average population size of all species in a representative sample, in accordance with the CBD 2010 indicators for species abundance (2010 Biodiversity Indicator Partnership). The index algorithm uses only original species in the characterization of regions, so as to avoid the quantification of opportunistic species, which may mask the loss of original species. Thus, MSA is the combined result of monitoring and modelling.

This index characterizes the integrity of biodiversity, seen as the mean abundance of original species relative to their abundance in undisturbed ecosystems.

By comparing biomes, as shown in Figure 1, it may be conjectured that the areas which will potentially suffer the heaviest loss of biodiversity in the future are likely to be savannas or tropical grasslands, as well as temperate and steppe grasslands.

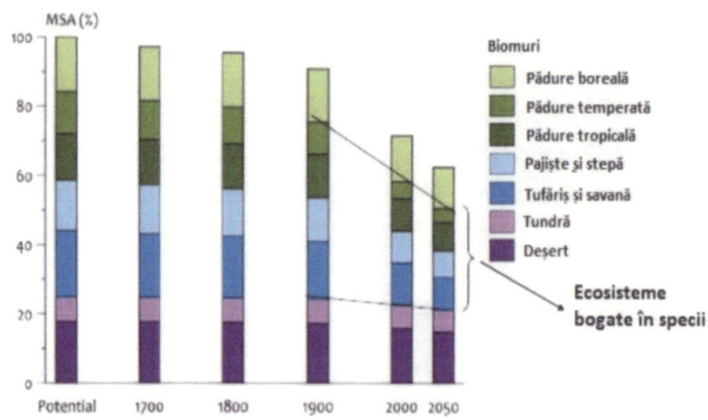

Figure 1. Evolutionary dynamics of extant global biodiversity (MSA %) (Brink, 2010)

3. Gap analysis: A useful tool for the assessment and development of perspectives on biodiversity

Aiming at relevant gaps in various biodiversity components the Gap analysis is not a substitute for in-depth biological inventories, but can be used as preliminary planning tool, serving as a basis for further study work, see [9], and for the decision making of institutions dependent on sound scientific data for managing natural resources ,see [10]. Initially defined as an assessment of gaps in the field of conservation in [11], the scope of this commonly applied method is extended within BeNatur project framework.

3.1. Assessment of gaps

A comprehensive gap analysis questionnaire was the main instrument for the assessment of gaps between partner countries, see [12] The appropriate completion was of high significance, as the outcomes are preconditions and necessary elements for further project steps. Complemented by best and bad practice examples, the data and additional documents gathered by each individual project partner allow useful insights in the actual management and implementation situation. The results are going to serve as a basis for the Joint Transnational Strategy and for the development of tools for the improved management and implementation of Natura 2000 sites in all partner countries. Basically, the questionnaire includes four parts, covering the following main points of interest:

1. legal procedures for the application of directives
2. management and organizational structures
3. ecological assessment and
4. socioeconomic assessment.

3.2. Outcomes and perspectives

In total, data of 11 questionnaires form 7 countries have been analyzed (Austria (1), Bulgaria (1), Greece (1), Hungary (2), Italy (3), Romania (2), Serbia (1)).

3.2.1. Assessment on legal implementation

In all countries (except Serbia), the legal implementation process is done, but in many cases deficits are still distinguishable, see [13-19]. In Italy and Austria the implementation is done at the level of federal states or provinces respectively, while in the other countries this is done at national level. The spatial dimension of the designated sites is defined in the laws by list of parcels (except Romania) and maps in the scale from 1:5.000 to 1:25.000. In the Habitats Directive no standard for the resolution of distribution maps of habitats or species is given, but the more accurate these maps are, the bigger is the benefit for further investigation and protection efforts.

In many countries, Natura 2000 is also affecting the laws of "Hunting and fishery" and "Spatial planning". Slightly more than the half of partners reported that Natura 2000 is also implemented in the laws concerning "Agriculture and forestry" or "Water management". In Austria Natura 2000 legislation is also implemented in several other thematic laws on the level of provinces (e.g. rural development, waste management, environmental protection).

Handling Impact Assessments (IA) of plans and projects at national level is the responsibility of the Ministry of Environment or a similar governmental department in most cases. Focusing on local responsibilities, there are only authorities in Austria, Romania, Greece and Veneto Region in Italy. The integration of local stakeholders as part of the governance system is still established insufficiently in most countries. The reported numbers of IAs according to Article 6 of the Habitat's directive are very heterogeneous, so there is no clear pattern to be observed. Only few partners where able to provide data on this question. The number of IAs and their results would actually be very interesting for the Natura 2000 network at EU level, as they are good indicators on threats to individual sites and could help developing reference decisions as well as a proper methodology for certain impacts.

3.2.2. Assessment on site management

There appears to be a significant lack of responsible managers in charge of the Natura 2000 sites. According to [13] in many countries, one person or institution is responsible for the management of several sites [13]. Therefore, the needed presence at the site level to be aware of ongoing processes at the site per se and to be at disposal for stakeholder's discussions is not given.

In only about 20 % of Natura 2000 sites there is a management plan available, taking most of the partner regions into account. According to [14] only Austria's number of management plans for Natura 2000 sites is significantly higher with about 70 % [14]. An additional deficit is the absence of mechanisms to approve the effectiveness of management plans. Only Italy, Romania and Bulgaria do have such structures, but even in those cases, the needed actions to improve site selection or management are missing, see [15].

Comparing the different degrees of involvement between the countries, it is apparent that the involvement of the stakeholder groups is the strongest in Italy, and in Greece and Austria this aspect also plays a certain, but minor role. There was good involvement of NGOs in Hungary in the designation period of Natura 2000 sites, but involvement has decreased in the ongoing management of the sites according to [16]. In other countries analyzed participative strategies and stakeholders involvement is rather unusual.

There are stakeholder groups that are obviously stronger involved than others. Farmers, hunters and foresters have a big influence on species composition and biodiversity of habitats due to their activities, which turns them into indispensable partners for a sound management of protected areas. Unfortunately, assessments on the degree of stakeholder acceptance are very rare throughout EU.

There is comprehensive information material about all aspects of the Natura 2000 sites in all EU-member states available on various levels. Online information is most frequently available, whereas printed materials such as booklets or leaflets are rather underrepresented. Expensive approaches of visualisation such as special booklets interpretative trails are expensive and their existence depends on the actual funding situation.

General difficulties and goals correspond to the findings of the European Environmental Bureau in [13], referring to an inconsistent mapping without any "common direction" and a lack of financial means as well as human expertise to handle the implementation of Natura 2000 guidelines.

3.2.3. Ecological assessment

Most partners can rely on interpretation manuals containing a national specification of the habitats and species. Only in Greece such a national specification of the habitats and species of the Habitats Directive is missing. National monitoring concepts exist however the implementation in the countries is not satisfying. National programmes for the coordination of activities in Natura 2000 sites would ensure the favourable state of species and habitats considerably. In addition, national projects would help to develop concerted actions all over a country for specific species or habitats.

Regarding the outcomes of BeNatur analysis, there are quite few programmes and projects on the national level reported by the partners, referring to a huge gap. There might be several reasons for that:

- There are simply no current national programmes and projects available.
- There are hardly any programmes and projects on the national level, maybe due to the fact that nature conservation is not the duty of national administrations but in the responsibility of the administration on province level.
- The access to information on the national level is limited, suggesting that national programmes and projects have not been successfully communicated to the regions and responsible site managements.

The results in terms of projects on site level are characterized by a very high variety among the different partners (from zero to more than hundred); probably due to the different stages of implementation of the Natura 2000 network. In some countries, administration is still in the process of nominating and designating sites by national law or developing first management plans. In this phase, hardly any projects dealing with concrete protection measures are implemented (Austria, Greece, Bulgaria). Other countries or provinces are one step ahead, having their management plans completed and already started implementing protection measures.

3.2.4. Socio-economic assessment

Regarding the socioeconomic part the results show that the BeNatur partner countries are facing huge "informational gaps". The assessment of actual and the estimation of needed expenditures to ensure an effective management of the sites was difficult or impossible for a significant number of project partners. This outcome suggests that the institutional as well as the human resource capacities of those authorities dealing with the management and planning of Natura 2000 sites are still not that strong as desirable. Partner regions indicated that the funding base for Natura 2000 network is very limited, regardless of the broad range of potentially available financial for the sites. The application and management of funds requires firm capacities, which appear to be insufficient in partner regions, referring to deficits at the institutional level. EU as well as national / regional funds play a key role. Private financing is very limited for Natura 2000 sites.

3.2.5. Best and bad practices

To sum up the best practices part of the analysis, it is apparent that especially the LIFE project framework comprises efficient project management, viable monitoring methods and beneficial project outputs in most of the cases. Thus, valuable experiences, creation of useful data bases and knowledge exchange are ensured. In addition, bad practices provide various lessons to be learned. According to the examples collected deficits in communicational activities, lack of adequate research and education structures in the field of biodiversity and nature preservation as well as the aspect of high administration costs are the main difficulties.

3.2.6. Conclusion

Gap analysis has proven to be a useful assessment tool within BeNatur project framework. As the results have highlighted, the current state of Natura 2000 management and implementation is heterogeneous and complex in the participating SEE countries. Manifold gaps have been identified in all fields assessed by the gap analysis questionnaire tool, requiring further analysis and interventions. Based on the gap analysis, important next steps will come into action to optimize implementation structures and management strategies of Natura 2000 sites.

4. Management of nature protection in Hungary

4.1. Biodiversity preservation

The National Biodiversity Strategy sets the general objectives for biodiversity conservation as well as for sectoral integration related to land use, water management, agriculture, forest management, fisheries, tourism, hunting and mining. In line with the objectives related to agriculture the National Rural Development Strategy was approved by the Government in 2012. In order to assist and complement in situ species conservation activities the so-called Pannon Seed Bank project was initiated in Hungary for the long-term ex situ conservation of the Pannonian biogeographical region's flora. This unique project establishes a joint seed bank for the agricultural and wild flora, meaning that the genetic diversity of the wild species as well as plants serving human nutrition are aimed to be conserved at one place.

4.2. Species conservation

According to our present knowledge, in Hungary there are approximately 600 moss species, 2,200 vascular plant species, 2,500 mushroom species and 42,000 animal species. The protection of flora and fauna is one of the important areas of nature conservation. Recently, new species to the flora and fauna of the country, or even to science have been discovered in Hungary (e.g. Hammarbya paludosa, Nepeta parviflora, Epipactis tallosii as well as the Vojvodina mole rat (Nannospalax (leucodon) montanosyrmiensis). Numerous new sites have been discovered for already known protected plant species. The population or habitat of some plant species of outstanding importance for nature conservation including e.g. Dolomitic Flax and Dianthus diutinus have been stabilised thanks to EU-supported projects.

At the moment there are 997 protected animal species in Hungary, of which 137 are strictly protected, while there are 720 species of protected plants, among them 71 strictly protected. The population of certain threatened species has been stabilised or even increased thanks to focussed conservation measures. For example, the Hungarian populations of Imperial Eagle, Saker Falcon, Red-footed Falcon and Great Bustard have increased significantly thanks to EU-supported LIFE projects. The ex situ breeding of Danubian Meadow Viper has proved to be very successful and natural populations are now regularly re-inforced with artificially bred specimens. Future priorities include site designation for the Vojvodina mole rat (Nannospalax (leucodon) montanosyrmiensis), as well as further genetic study of the mole rat species.

4.3. Agriculture and nature conservation

Hungary can be considered as an agricultural country not only with regard to its history, but also to its present land use figures. The great majority of our natural values is linked to ecosystems created and maintained by agricultural activities, which is well represented by the fact that out of the 846,537 hectares of protected areas 26% is grassland, and 12% is arable land. The same parameters for the 1.99 million hectares of Natura 2000 sites are 25% and 27%. Sustainable farming on salt steppes, wet meadows, wetlands and forested pastures is also a precondition for preserving the unique natural values of the Pannonian biographic region. Grazing and mowing methods, pesticides, fertilization and crop rotation all determine whether the given farmland can fulfill its role as nesting, feeding and living habitat for a great variety of protected species. The dramatic decrease in grazing animal population in the last few decades – which indirectly led to the decrease of grasslands as well – means a serious challenge for nature protection, despite the great richness of natural values preserved to this day. In connection with farming practices, some of the main objectives are to prevent the spread of invasive species and to preserve tree lines, hedges, green infrastructure and temporarily flooded wetland habitats. Hungary is in a special situation as 289,383 hectares of the protected areas (34%) is in state hands and managed by national park directorates, in the case of agricultural areas mainly through the involvement of local farmers. Through national park directorates nature protection plays a crucial role in the preservation of protected traditional breeds of livestock, which are fundamental from the perspective of high nature value farming systems as well. In addition to the strict legal regulation and ownership structure (72% of protected areas is owned by the state), Hungary can successfully exploit opportunities offered by the rural development funds. The compensation of farmers' obligations, and the success of voluntary programmes introduced in order to develop habitats resulted not only in increasing public awareness, but in the positive population trends of some target species (i.e. Great Bustard). Within the framework of the agri-environmental programmes of High Nature Value Farmland Areas currently more than 2,200 farmers fulfill the land use commitments voluntarily undertaken for nature protection purposes, altogether on 204 thousand hectares. As a compensation for the obligations of farmers utilizing Natura 2000 grasslands, in 2011 area-based payments were distributed covering more than 250,000 hectares.

4.4. Management plans of protected natural areas

The basic elements of Hungarian conservation planning system are the management plans of protected natural areas. While the National Conservation Plan refers to the whole territory of Hungary during a five-year period and regional conservation plans ("landscape protection plans") cover the territory of several counties, the management plans of individual protected sites apply to areas from 1 ha to 80,000 ha. The period of validity is ten years. Conservation management activities are defined by the Act on Nature Conservation (1996) as surveying, registration, protection, guarding, maintenance, interpretation and restoration of protected natural values and areas. According to the Act on Nature Conservation every protected area has to have a management plan which is part of conservation legislation, as management plans are given out in a ministerial decree by the minister responsible for nature conservation. New protected areas can be designated only with complete management plans. The management plan documentation consists of three main parts: the supporting documentation, which contains the physical, economical features, the conservation status of the protected area and the conservation objectives; the detailed management plan with the conservation strategies and detailed description of conservation measures, restrictions and prohibitions; and a short version of the detailed management plan which is part of the ministerial decree mentioned above. The table of content and the preparation (the process of conciliation on local level etc.) of the third part is determined by a ministerial decree, while the requirements of the first and second part are issued in a ministerial ordinance. Preparation (and realization) of management plans of protected areas is one of the most important tasks of the ten national park directorates in Hungary. At the present time there are 211 protected areas (national parks, landscape protection areas, protected areas and natural monuments) in Hungary, among them 73 protected areas have a management plan issued in a ministerial decree.

4.5. Ecotourism in protected natural areas

Interpretation of natural and cultural values in protected natural areas is primarily attained by ecotourism, with respect to local conditions reducing the environmental effects to a minimum. Approximately one-tenth of Hungary's territory is classified as a protected natural area. There are ten National Parks, 38 Protected Landscape Areas and several Nature Conservation Areas: they hold dense hillside forests, rolling hills, wild rivers, wide plains and areas where limestone rocks and caves conceal amazing arrays of hidden treasures. In these beautiful protected areas there are 27 visitor centers providing special information about the sites, including interactive exhibitions on the natural and cultural heritage, also considering the achievements reached by nature conservation.

Environmental education is a high priority question in ecotourism. Its most important goal is to enhance a positive attitude towards nature conservation among the youth, to bring nature closer to them, and helps to get them acquainted with its values. The Ministry of Rural Development, responsible for nature conservation and tourism, plays an important role in organizing ecotourism activities in protected areas with co-operation of national park directorates. The federal ministry annually organizes the Week of Hungarian National Parks with a workshop on ecotourism and a wide range of ecotourism programs for the public. In 2012, the workshop will focus on the (eco)tourism of World Heritage Sites. We also organize The Heritage Interpretation Sites of the Year competition in two categories (visitor centers and nature trails) with co-operation of the ministry responsible for tourism and the state-owned company responsible for tourism marketing.

4.6. Geological values of protected natural areas

Geological objects have been regarded as part of our natural heritage from the very first attempts at the comprehensive regulation of nature conservation.

The Act on Nature Conservation provides the possibility for protection of practically the whole range of elements of geoheritage. Under the Act, protected status shall be given to: geological formations and key sections, minerals, mineral associations and fossils; important sites of protected minerals or fossils; superficial or geomorphologic formations and the ground surface above caves; typical and rare soil profiles; which deserve such protection out of scientific, cultural, aesthetic, educational, economic or other public interest. On the other hand the law broadened the scope of ex lege protection of caves to major springs, as well as to sinkholes; the latter two qualify as protected areas of national importance. Caves, minerals, mineral associations, fossils are defined as protected natural assets, and are in state ownership.

The first national inventory was made on the ex lege protected items. Inventory of caves was compiled in 1977, and up to now the National Cave-Cadastre contains 4,120 elements. Similar registrations for sinkholes and springs could only be started in 2002, with elaboration of the methodology. Up to now, 795 sinkholes and 2,732 springs have been registered.

Geoscientists have played a significant role, not only in laying out the concept of nature conservation, but in its implementation as well as by developing the facilities of protected geosites and launching the 'Key-section Programme'. As a result of further research, the Stratigraphic Commission of Hungary listed 485 surface key sections, however, it does not mean that all of the key-sections are protected.

From among the scientific societies providing an important contribution to geoheritage conservation, the Hungarian Geological Society, the Hungarian Geographical Society, the Hungarian Karst and Speleological Society and the Hungarian ProGEO Association should be mentioned.

5. Túrkeve landscape development program as best practice example for the reintegration of unemployed people

In the focal point of the Túrkeve Landscape Development Program stands the man who lives together with the countryside and landscape; the local community and its changes, which fundamentally define the development of the natural values.

In 1995 after the regional nature conservation committee's advice, the association at first time had the opportunity to establish a wider and regional organisation. One of the biggest social problems of the region is the high rate of the unemployment, the lack of job opportunities for less educated people, minority and elders and the migration among the youth which is caused by the narrow living possibilities.

The project tries to find new social solutions as the previous experiences show that recent approaches use 'end of the pipe solutions' rather than cure the root of the problems. In our vision the cause of the problems should be found in the improper landscape management. The program combines the sustainable use of the local resources with the local human background which is completely new in this structure. Farming on 300 ha with 27 horses, 17 cattle and approximately 100 sheep considering the viewpoints of the nature-friendly grassland management and traditional grazing methodology we can provide job opportunity for 20-30 people in season which is equal to a several thousand ha' intensive farm's capacity. The income provided by the farm contribute to the development of the region, to the decrease of the poverty and at the same time the degradation of the natural resources is missing, the development happens in a sustainable way. A farm which is managed according to this philosophy can count in a long-term as it ensures the living of the local people throughout generations without the degradation of the arable lands.

The emphasized target groups of the project are local and regional people with lower academic levels, old and unemployed people and the members of the minorities whose job opportunities are limited due to the negative prejudices and their decreased abilities. Indirectly those young people are also our target group who cannot see their future ensured in Túrkeve or in the region. For the poorest individuals accommodation in the farm buildings is provided, furthermore trainings for the workers who apply for a job are realised. Moreover we are continuously apply to the town's communal work program, occasionally we can hire 3-5 persons for several months or even for a year and it is nor rear that from them someone will get a permanent job at our association. Within the frame of the compulsory volunteer program – which is necessary for the social security benefit – we host men and women almost as an exception in the town.

The project gives long-term benefit for the involved target group as the farm gives occupation throughout the year and provides ensured living for them. At the same time they keep the functional capacity of the ecological services, treat the land, graze the animals sustainably and do not exploit the natural resources, moreover, thanks to the re-establishment of the traditional farming methodology the flora and fauna of the area has been growing, rare and locally extinct species appeared again. Seeing the results of their work they feel the town and landscape as their own, which raises linking and keeps the youth here. The permanent income often makes people more open, self-confident, gives them enthusiasm and persistence. They can use the obtained knowledge elsewhere, in their own farm or at their future employee where they can count the time spent in our association as an experience. Farmers who agreed with the association's aims and ambitious let their land by lease which helps them concerning the management of their land. More and more inhabitants, seeing the aims and goals of the effort, let by lease, sell their land or farm in agreement with the association.

In the future we would like to improve our possibilities with further areal expansion and ecotouristical investments. Our most important plan is the improvement of the employment rate. Employment programs have started with touristic improvements, trading activities based on local products and developments such areas of work which are independent from educational level (e.g. growing herbs, proceeding fruits, horticulture). We are trying to find small-scale traders and provide them platform where they would be able to sell their products easier. Firstly it has a positive effect on the standard of living, secondly it makes stronger the identity, thirdly it pretends the small plots' divers habitat structure.

With the area we continuously apply for European Union and national funds which covers the expansion of the employment possibilities and contribute to the ecotouristical developments. At the moment the organisation has an approved LIFE Nature project which started in September 2012 and will last until December 2014. In the frame of this initiative we are going to renovate the existing farm house, purchase animals such as Furioso North-Star horses, Hungarian cattle, racka sheep which are all Hungarian endemic breeds, prepare the Natura 2000 management plan for the site, establish a nature friendly water management in the area. Nevertheless one of the most important objectives of the project is to collect the owners of the area and try to organise a board of landowners in which the farmers vote together and manage their land in cooperation with each other.

According to the organisational aspects, one person leads the farm theoretically, direct the activities out of the fields. One member of the staff maintains the administrative side of the farm, apply for funds, organise the selling of the offspring, provide the financial background for the investments. The third pillar of the organisational graph is the practical coordinator who divides the tasks among the workers and check whether the exercise has been done.

The program combines the social and nature conservation approaches as today's sectoral thinking tries to solve environmental and social problems with several branches which have different and most of the times opposite effects from each other.

In the protected area the ruling principle is the reserve-theory: the state maintains the area and rule out the local community from there. The labour market desperately tries to grab such sectors which are harmful for nature or do not serve sustainability, based on external resources and finally ends with the destruction of our own natural resources whilst the poverty is growing and the 'social scissors' is getting wider. This is why we consider important to harmonize the landscape features with the human activity which partly makes the living independent from the changes of the economy as self-sufficiency means independence. As the area which is farmed is under EU protection it is not threatened by the risk that the long-term use of the land would change. Furthermore the treated land belongs to the association or is let by lease with long-term contracts. The permanent employment is ensured as the land is managed by the association. The livestock is also in the organisation's hand. There is a good relationship with the other farmers as well with those who maintain their land in line with the objectives.

Figure 2. Traditional horseman from Túrkeve

Also for Austria agriculture plays an important rule in Natura 2000 areas. Many agricultural sites, especially in peripheral grassland areas like for example the Enns Valley, are directly located in Natura 2000 areas where often conflict potentials between land owners and nature protection arise. State funds should serve as first approach to solve the problems and promote extensive agriculture.

6. Agriculture and nature protection in Austria – Interventions as example for Natura 2000 funding

In 2008 the Austrian Court of Audit remarked that only a part of the necessary ordinances of Natura 2000 management were put into force. Management plans were available in different scope, content and quality and not legally binding. The list of recommendations includes clear priorities and implementation in guidelines. A comprehensive supervision and monitoring system of the protected sites under the aspect of long-term financing and available resources should be installed, see [20].

In Austria there are 220 identified Natura 2000 sites, which account for 14.7 percent of the state territory. At least 148 of them have also been legally enacted. The majority of the sites have been enacted according to both the Habitats Directive and the Birds Directive. Due to the federal character of Austria, there are nine different Province laws about nature protection in each federal state, but there is no national, common law. According to [21] in total, about 50 percent of the areas in Austria are used for agricultural purposes [21].

However, intensive agricultural usage of the areas is often not compatible with the objectives of nature conservation. The potential for conflict arose due to both the expulsion of the protected areas without involvement of the land manager as well as the provisions of nature conservation laws for a careful subsistence strategy for the conservation and improvement of the protection objectives in accordance with the FFH- and Bird Directive. From the agricultural viewpoint, nature conservation often represents a management difficulty. The various interests between land use and conservation constraints and the associated loss of earnings are the cause. The lack of information and involvement of all stakeholders in the implementation process of the EU regulations through a top-down process and the contradictory recommendations and advice of their own interest groups further exacerbate the conflict.

Solutions are being sought out with the objective of creating a multifunctional agriculture in which farmers function simultaneously as conservationists and sustainers of the cultivated landscape. Various assistance programmes are being offered for the implementation of the Natura 2000 goals in Austria, which can be quite lucrative for the land managers. Depending on the programme, desired achievements of the farmer (e.g. landscaped structured in small sections, the disposition of deadwood, biodiversity, etc.) that can positively affect the conservation status of the area are financed.

6.1. Austrian programme for environmentally friendly agriculture (ÖPUL)

In reference to [22] with the Agricultural-environmental programme [22], the Austrian Programme for the advancement of environmentally friendly, extensive and habitat-protecting agriculture, an environmentally sound management of agricultural areas is championed. Included in the more detailed goals are the advancement of environmentally friendly agriculture (and pasture farming of low intensity), the preservation of traditional and especially valuable agriculturally used cultivated landscapes, the conservation of the landscape, the advancement of the inclusion of environmental planning in agricultural practise, payment for the realisation of national and societal agricultural and environmental policies through the advancement of contractual nature conservation and measures to protect of waterways, soil, and groundwater, in addition to the advancement of organic subsistence strategies and the securing of suitable compensation for the offered environmental services.

Compared with some other EU countries, which offer their environmental programmes only in marked-off, environmentally sensitive areas, an integral, horizontal approach was chosen for ÖPUL. The legal basis is a special guideline in which general and provision-specific eligibility criteria are set. In 2011, € 549.2 million was paid to 114,508 enterprises for ÖPUL services over 2.2 million hectares. 114,508 enterprises, representing 74 per cent of all agricultural enterprises and 89 per cent of agriculturally used areas, participated in ÖPUL. With this high rate of participation in an environmental programme, Austria lies in the forefront of the EU countries. The areas contained in ÖPUL (excluding alpine pastures) account for about 2.20 million hectares. The average aid per ÖPUL enterprise was € 4,795.

6.2. Biotop conservation and aid programm (BEP)

Contractual nature conservation lies at the forefront of protection of species and habitats. Landowners work on a voluntary basis with environmental protection authorities. In places where the stipulations of nature protection laws cannot be implemented, many cultivated landscapes are in danger and on the decline. The conventional usage is no longer economical, the areas become increasingly overgrown, and biodiversity disappears. In order to keep these ecologically valuable areas safe, the land managers receive aid from the provincial government. People should be made aware of the ecological value of the meadows and pastures. In 1987, the Styrian Province Government, working together with the chambers of agriculture, already implemented the Biotop Conservation and Aid Programme as a specific conservation aid programme. The primary goal is the conservation of valuable, extensively farmed grasslands. They are no longer fertilised and are only mowed one to two times per year. The cutoff of yield is compensated for only following certain regulations. With this programme, habitats diverse in species are preserved as the last remnants of the original flora and fauna and a progressive depletion of the natural features of the agricultural production area is simultaneously counteracted.

6.3. Austrian programme for the development of rural areas 2007-2013

Objectives for the conservation and improvement of the rural legacy – protection of nature is the conservation and development of resources valuable to the field of environmental protection and the regional uniqueness of the cultivated landscapes, especially that of habitats and species, supporting local stakeholders in order to take advantage the full potential for natural spaces in society. The preservation of region-specific landscape qualities is understood to be a service to society; good conditions for value added afford the development of competences for management of natural areas by providing services in nature conservation. The establishment of national parks or nature- and biosphere reserves are examples of sustainable development. Projects for protection and development of biotopes for the conservation and development of habitats and protected species are financed, including those for land restoration projects for valuable wetland habitats as well as the creation and conservation of landscape structures, e.g. dry stone walls. Even the costs of land acquisition and the management and support of the protected sites are covered.

These aid programmes are partially tied with great bureaucratic costs, and success is not always ascertainable. Moreover, some programmes run until 2013 and will be redrawn after evaluation and to conform with the new Common Agricultural Policy.

Further important for the conservation and the management of nature protection sites is the realization of national and international cooperation projects. For Austria the LIFE Project in the National Park Gesäuse was a significant impulse for nature protection measures in the Natura 2000 sites "Ennstaler Alpen – Gesäuse" and "Pürgschachen Moos und ennsnahe Bereiche zwischen Selzthal und Gesäuseeingang". Through collaboration of different institutions dealing with nature protection a significant improvement was able to be reached in both sites.

Figure 3. The Natura 2000 site "Ennsaltarme bei Niedersuttern" in the Styrian Enns Valley is managed after ÖPUL guidelines.

7. LIFE project: Conservation strategies for woodlands and rivers in the Gesäuse National Park

The EU's LIFE Nature Programme is a funding instrument primarily dedicated to the implementation of the Flora-Fauna-Habitat guidelines. It deals with the establishment of the NATURA 2000 nature conservation network in the EU and also concretely the introduction of provisions for the protection and conservation of endangered and protected species and habitats. Even in the new financial period from 2014 until 2020, the funding programme is to be strengthened in line with environmental, nature and biodiversity protection projects within the member states. Within the Gesäuse National Park an important Life Project for conservation of woodland and natural rivers was implemented.

The Gesäuse National Park was founded in 2002 and is located in the Alps of the Enns valley. Approximately 50 per cent of the region is covered by forest, predominantly by spruce, fir, and beech. According to [23] the national park serves as a primary habitat for many species important to conservationists [23]. The approval and the reestablishment of the natural dynamic, predominantly the river dynamic of the Enns River and the Johnsbach, the largest feeders to the Enns River in the Gesäuse, are counted among the most important tasks of nature protection. In the Natura 2000 sites "Ennstal Alps – Gesäuse" and "Pürgschachen Moos and areas between Selzthal and Gesäuseeingang close to the Enns river" important measures for the conservation and protection of habitats and species were implemented in accordance with the LIFE Project. The objective was to improve and upgrade the habitats for target species as well as target habitats along the waters of the rivers Enns and Johnsbach. Measures were implemented for the bordering mountain forests and alpine pastures in the form of management plans. The spruce forests (foreign to the area) were converted into predominantly hardwood mixed forests.

Composition of sectoral management plans:

- Management plan Johnsbach, see [24]
- Forest management plan, see [25]
- Alpine pasture management plan, see [26]
- Visitor direction concept, see [27]
- Neophyte management plan, see [28]
- Management for debris accumulating in the area

Furthermore, the population was also integrated in the project (e. g. surveys about Enns guidelines, see [29], as well as participation processes, workshops, excursions). Another form of contribution was also the extensive public relations work (events, informative materials, information boards, educational trails, visitor direction concepts).

The most meaningful goal was the restoration of an ecologically functional continuum of waterways of the Enns and the connection to its main feeders, Johnsbach and Palten. The deconstruction causes the reintegration of floodplain regions and aids the natural habitats such as areas covered by pioneer vegetation and Alnetum incanae according to [24]. The target species along the Enns and Johnsbach are Lutra lutra, Eudontomyzon mariae, Cottus gobio, Leuciscus souffia agassizi, and Alcedo atthis. Through habitat conservation measures on alpine pastures, the habitats for Tetrao tetrix and amphibians (Bombina variegate) were able to be enhanced and expanded. The cooperation with neighbouring natural forests (Kalkalpen National Park, Dürrenstein Wilderness Area) should be strengthened. An ongoing evaluation of the measures serves to safeguard the goal standards in the course of the project. The measures for protection against natural hazards along the Johnsbach can be fit to the new ecological demands and be optimised for the target species (EU Water Framework Directive). The cooperating partners of the project were the Department of Flood Protection of the Styrian Province Government, the Austrian Service for Torrent and Avalanche Control, district office Enns- and Paltental.

7.1. Results and outlook

In reference to [30] the following data should briefly summarise the results of the project [30]:

Regarding the Enns, improvements of the ecological state of a total of 105 kilometres of the river are planned (implementation EU Water Framework Directive). Between the Enns and Palten, an area of five hectares was restored and left to its own natural dynamic. In the area of the mouth of the Palten (into the Enns), about one kilometre of the river was restored. About five kilometres of new pioneer habitats were created on the torrent Johnsbach, which was also once again made passable for fish. The forest management plan encompasses a total area of 5,500 hectares. Until now, over 300 hectares of spruce forest was thinned out for the development of mixed forest. The pasturing concept for the alpine pastures encompasses 700 hectares in total and will be realised in steps. The already realised provisions were confirmed by the supplementary monitoring, which will continue to be conducted in the future. The new LIFE+ project "Flusslandschaft Enns" ("River landscape Enns") of the Styrian Province government puts the restoration of the Enns in motion. The Gesäuse National Park has already taken further steps towards a networking of biotopes in the region Northern Kalkalpen/Eisenwurzen through cooperation on the Alpine Space Project "ECONNECT" , see [31], with partners National Park Upper Austrian Kalkalpen and the Wilderness Area Dürrenstein. Main goal is the development of a region for nature and culture for the benefit of its residents and visitors.

Figure 4. Johnsbach River in the Gesäuse National Park

Another important project for conservation of habitat and species is realized in Italy with the tourist management in natural beaches, where the illegal frequentation for bathing of natural protected seashores affect the conservation of the fragile ecosystem of beach and dunes and the preservation of rare species such as Little Tern and Kentish Plover. After fifty years of tolerance the situation became deep-rooted and the tourist management is now the main threat for the most important natural seashore of Emilia-Romagna region. The Province of Ravenna is trying to face the problem by new specific rules and by a project of sustainable development in the beach, to turn the commitment in a resource.

8. Tourist management in natural beaches in Emilia-Romagna in Italy

The Emilia-Romagna region has 150 kilometers of seashore along the north Adriatic Sea, from the Po di Goro mouth (Ferrara) to Riccione (Rimini). Only 15 kilometers (10%) are protected as Natural Reserves and are now included in the Natural Park of the Po Delta, but are however used for bathing tourism, in spite of the no entry in force. The most important natural beach in the province of Ravenna is called "Stream Bevano Mouth Beach", because it is around the last natural river mouth of the North Adriatic Sea. This beach is long about 6 kilometers (3 km north and 3 km south of the river mouth) and there are two small towns with bathing beaches at the north and south ends. The beach has been protected since the 1979, but the bathing tourism continued undeterred till few years ago, when started the activity here described. Stream Bevano Mouth Beach is very important for nature conservation because there are very good examples of some rare habitats protected by the directive 92/43/EEC, such as 1210 "Annual vegetation of drift lines" (locally characterized by the Salsolo kali-Cakiletum maritimae Costa e Manzanet 1981); 1320 "Spartina swards Spartinion maritimae", locally characterized by the endemic Limonio narbonensis-Spartinietum maritimae Pignatti 1966 - Beeft. & Géhu 1973); 2110 "Embryonic shifting dunes" (locally characterized by the Echinophoro spinosae-Elymetum farcti Géhu 1987); 2120 "Shifting dunes along the shoreline with Ammophila arenaria" (locally characterized by the Echinophoro spinosae-Ammophiletum australis Br.-Bl. 1933); 2160 "Dunes with Hippophae rhamnoides" (locally characterized by the endemic Junipero communis-Hippophaetum fluviatilis Géhu & Scoppola in Géhu et al. 1984), according to [32]. Bathing tourism is a heavy threat for the conservation of habitats and species, because people walk over the dunes, trampling upon the fragile sand-specific vegetation and because they scare birds, stopping their nesting in the breeding season or stopping over during the migration, all over the World, see [33, 34].

The bathing tourism is devastating for beach and dune ecosystem and the situation in the "Stream Bevano Mouth Beach" was more and more deteriorating, till the finally death blow caused by the mechanical cleaning, started in the middle '90 by the local Municipality, unbelievably asked by the unauthorized tourists. Both the high disturbance of the surplus of people standing on the beach each day from late April to the beginning of October and the weekly transit of large tractors for mechanical cleaning caused the destruction of habitats 1210 "Annual vegetation of drift lines" and 2110 "Embryonic shifting dunes", the erosion of 2120 "Shifting dunes along the shoreline with Ammophila arenaria" and caused also the local extinction of some nesting birds protected by the directive 09/147/EU, many years ago, such as for the Stone Curlew Burhinus oedicnemus that nested last time in 1949 according to [35] or during the last years, for Little Tern Sternula albifrons (last nesting in the middle '90) and Tawny Pipit Anthus campestris (last nesting in 2000) or caused the remarkable reduction of Kentish Plover Charadrius alexandrinus, passed from about 30 pairs in the '80, to 8-10 in 2004, 14-18 in 2005 and 0 in 2006 in reference to [36]. Moreover, the beach cleaning unbalanced the ecosystem, collecting the fundamental organic substance settled by the sea and breaking the food chain at its base. The Province of Ravenna is responsible for the Land Plan of this part of the Park of the Po Delta since 1988. The procedure started in 1991, without an office in charge of parks management. The first Land Plan was rejected by the Region Emilia-Romagna in 1997, because it was deficient about nature conservation. The Province of Ravenna set up the Parks Office in 2000 and the new Land Plan had been elaborated in 2004, adopted in 2006 and now, finally approved by the Region in April 2012. The Land Plan is about a wider area, called "Classe Pinewood and Cervia Saltpans", including large pinewoods and some brackish marshes and saltpans. The natural beach is part of a complex with the native ecological succession from the foreshore to the inland woodlands, with humid dune slacks, grey dunes, lagoons. This plan faces the problem of the unlawful bathing tourism and tries to solve it, considering its endorse by the local municipality and the long lasting custom of tourists. First, the Land Plan strictly orders protection of this precious beach: each change in morphology and hydrology of dunes and beaches; any kind of human construction; pits and dumps; hunting, fishing or other kind of animals trouble; plants damaging; road traffic; flying over; camping; trampling in dunes; dogs introduction; lightings. Here are only allowed: the hand cleaning of beach from man-made waste; possible rebuilding of dunes using natural engineering. The beach is divided in three zones, with different levels of allowed access. The first 750 meters close to the north and south small towns are "beach.c" and here are allowed also the mechanical cleaning, the access and the bathing tourism. The two central parts, among the towns and the river mouth, each of 1.5 kilometres, are "beach.b" and here are forbidden the access from April, 1 to July, 15 (to preserve nesting birds), the mechanical cleaning, the sea-settled woods removal and are allowed the access and bathing tourism from July, 15 to March, 31. The inner zone of 1.5 kilometres around the natural river mouth is the "beach.a" and is strictly protected also about human presence: the access to the beach is completely forbidden all over the year and the same is for beach cleaning.

Figure 5. The habitat 2110 "Embryonic shifting dunes" (locally characterized by the Echinophoro spinosae-Elymetum farcti Géhu 1987)

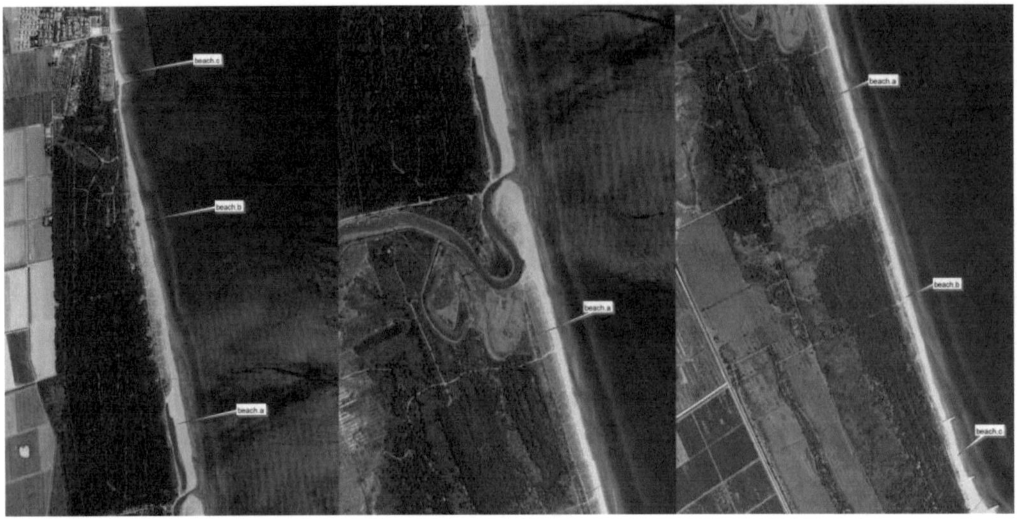

Figure 6. The beach zoning from North to South

To make the plan acceptable to the local community and tourists, this project for sustainable tourism development of the accessible beach was processed. The objective is to establish a "Park's Beach" in the first 750 meters close to the small towns and, after the July, 15 also in the two parts of 1.5 kilometres in the middle part of the protected seacoast and a "Wild Seashore" in the beach around the Stream Bevano Mouth (1.5 kilometres). The "Park's Beach" should be a bathing beach different from the other "normal" and very artificial beach of the Emilia-Romagna coast, that everywhere have flattened sand, fixed beach umbrellas, bathing resorts. This beach should have dunes at the back, irregular sand, more presence of organic substance settled by the sea (from the first 750 meters regularly cleaned to the central 1.5 kilometers where only human waste should be collected) and should be free from fixed beach umbrellas and bathing resorts. Only a small and temporary building at the entrance of the beach is provided for toilet facilities and in order to rent traditional beach tents and to sell cold drinks and ice creams. Also the introduction of a controlled number of people (maximum 300 people per day) aids to create the impression of a special beach and is very important to preserve the ecosystem. The "Wild Seashore" is even more natural, it is not at all bathing beach, but a completely natural ecosystem where the only aim is to preserve wildlife and where man is just a watcher. The visit of the area is only possible by guided tours. The beach should also be interesting from a cultural point of view, because it should be a real example of the local seashore landscape and habitats before the bathing tourism boom of the Fifties. At the end, both the kind of beach could become an economic resource; the "Park's Beach" by the introduction of an entrance ticket and thanks to the rentals and sales (with a 3.00 euro entrance fee the assessed annual gain could be about 54,000.00 euro); the "Wild Seashore" organizing the guided hikes to watch the natural landscape and habitats and to observe the wildlife of a unique kind of beach. Together with these new rules, the Corpo Forestale dello Stato (National Forest Rangers) and the AsOER (NGO Association of Ornithologist of Emilia-Romagna) started a campaign of information, to explain to bathing tourist and local people the importance of dune habitats and of some birds nesting on the beach and why it's vital to let some parts of the seashore free from human presence; the AsOER also started in 2008 a monitoring of the Kentish Plover nesting population.

8.1. Conclusion

The Plan was adopted by the Province of Ravenna in 2006, but a real control of the people along the protected beach started coyly in 2010, together with the interruption of mechanical cleaning and only from 2012 the no entry is actually enforced, because the Region Emilia-Romagna conclusively approved the Land Plan of the Park last April, 2012.

The first results arrived immediately: the habitat 1210 "Annual vegetation of drift lines" and 2110 "Embryonic shifting dunes" immediately re-colonized the beach, by growing of Cakile maritima, Salsola kali, Euphorbia peplis, Medicago marina, Eryngium maritimum, Echinophora spinosa, Calystegia soldanella, Agropyron junceum ssp. mediterraneum (= Elymus farctus ssp. farctum), between the white dunes and the foreshore.

The nesting population of Kentish Plover started forthwith to increase (5-6 pairs during the 2011 breeding season) and the Little Tern is really now trying to settle in again with a new nesting colony (breeding rituals recorded) at the beginning of May 2012.

Ovcar-Kablar in Serbia has been defined as "landscape of outstanding features". In the protected area education and awareness rising activities concerning topics like environment, ecology or healthy lifestyle are of high importance. Regularly there are several programs, projects, seminars, workshops and campaigns for the ecological education of the population.

9. Environment protection in Serbia

Nature protection systems in Serbia, as well as the basic categorization of protected areas are defined by Law on Environmental Protection (2004, 2009) and by Law on Nature Conservation (2009). Categories of protected natural areas are in accordance with the categories regulated by the International Union for Conservation of Nature.

Total area of protected areas (classified into 7 categories) is 518,204 ha that is 5.86% of the territory of Serbia. One of area protection categories is Landscape of outstanding features. It is an area of distinguishable visual aspect with significant natural, bio-ecological, esthetic and cultural-historical values which has been developing over time as a result of interaction between nature, natural potentials and traditional way of life of local population (Law on Nature Conservation 2009, Article 33). Up to now, sixteen areas have been classified as Landscapes of outstanding features and Ovcar-Kablar gorge is one of them.

9.1. Ovcar-Kablar Gorge landscape of outstanding features

Ovcar-Kablar Gorge was declared a protected area of the category Landscape of outstanding features by the Regulation of Serbian Government in 2000, and protection measures were closely determined by the Act on Proclamation of Protected Area. Area under protection is

2,250 ha and Tourist Organization of Cacak was given the role of administrator. Primary natural feature of the gorge is landscape diversity, impressive Ovcar and Kablar massifs and West Morava, which separates them, and which has formed unique meanders that represent geomorphologic phenomenon known as "incised meanders". Gorge length is 20 km and it is deepest in the middle part where the depth, in relation to Kablar peak, is 620 m and in relation to Ovcar peak – 710 m. The gorge boasts a significant number of surface and underground karsts formations.

Diversity of landscape, geological substratum and land, climate of the area and microclimate of some gorge parts conditioned a large number of various habitats, thus richness of flora and fauna. There are terrestrial and aquatic ecosystems with distinctive flora and vegetation that are often of relict character. With regard to the total number of species registered in Serbian Flora, there are about 600 species registered in Ovcar-Kablar gorge, which is 19.2 % of plant species. About 160 bird, more than 30 mammal, 10 reptile and 21 fish species have been registered on the territory of Ovcar-Kablar gorge. Best examined fauna species on the territory of Ovcar-Kablar gorge are birds. Due to a large number of bird species, particularly a great number of internationally significant species, this area has IBA status.

9.2. Development of educational contests on the territory of the city of Cacak

In 2005, Development Strategy of the Cacak Municipality and Local Ecological Action Plan were accepted. Education of population and positive development of environmental awareness stand out as a significant aim in these strategic documents. Having these documents accepted, we started an intensive work on the development of educational contents on the territory of Cacak, in terms of both creating and increasing the variety of educational contents and on strengthening the institutional capacities needed for their realization.

Educational contents are realized on several levels and through several programs. Education of citizens, out of educational system is realized through informal forms of education which includes projects, seminars, workshops, actions, campaigns etc. These forms of educational programs have been developing for years and a number of institutions took part in their creation (Local government, schools and pre-school institutions, NGO, media etc.)

The most significant educational programs which have been continually realized on the territory of Cacak year after year are realized through Program for Environmental Protection Fund Usage of the City of Cacak. Particular item of the Program is related to ecological education of the population. Education of the population is carried out on two levels.

The first level of education is realized in pre-school institutions, with an aim to develop a concept of healthy living, positive attitude towards the environment, nature and life in accordance with natural environment. This program is predominantly aimed at generating ecological awareness and developing healthy lifestyle habits with pre-school children.

Ecological education of pre-school children is carried out in several steps and lasts for entire years. The first step is concerned with the education of kindergarten teachers who develop professionally for the work with children through accredited ecological seminars. After that, kindergarten teachers organize thematic workshops in their groups. Some of the topics of organized workshops have been: "Let's establish Ecological Code of Conduct", "How we pollute water, air, the Earth", "Let's save birds", "Recycling in the kindergarten", etc. Besides ecological workshops pre-school institutions also organize excursions into nature, ecological patrols and collection of recyclables.

The second level of education is realized for primary and secondary school students. Schools are given funds for the realization of educational projects through Program for Environmental Protection Fund Usage. Basically, each project is realized through several major activities. First activity includes seminars lasting for several days, organized for the groups of 20-25 students in Eco-educational Center Rosci located on the border of protected area. Seminars consist of workshops which can be divided into theoretical (organized in the camp facilities) and practical (organized in the field- in nature). After that, field activities are organized- practical work. Location for the practical work depends on the seminar topic. At the end, promotional materials are prepared and distributed as a part of campaign that is organized in a school and/or some public institution. During 2011, among others, the following thematic projects were realized: " Monitoring of birds' state in the protected area Landscape of outstanding features Ovcar-Kablar gorge", "Remove ragweed and tree of heaven" and "Small Petnica in Rosci".

Apart from educational seminars and workshops for pre-school and school children, several actions are also organized every year. Almost all schools and pre-school institutions from the territory of the City of Cacak as well as entire citizenship take part in these actions.

- Within the action "Let's Clean up Serbia", on the occasion of the World Environment Day, on June 5th, Ministry of Environment, Mining and Spatial Planning in cooperation with Local governments, organizes "Big Cleaning of Serbia" - a big cleaning up action on the RS territory. Over 70 % of local educational institutions take part in the campaign every year. Teachers and students of participating schools organize the cleaning up and/or collection of recyclables.
- On the territory of Cacak, on the occasion of the World Environment Day, School for Catering and Tourism organizes, has organized ECO FEST three years in a row. All schools from the city territory can take part in the FEST and the theme of the FEST is chosen by the organizer. The main event at the last year's ECO FEST was the proclamation of the best arranged school yard on the territory of Cacak.
- Mountaineering club "Kablar", has been organizing the action "Big Mountain Race" for 15 years, through Ovcar-Kablar gorge, for cyclists, marathon runners and walkers. The race is of national character and all "nature lovers" have the right to participate.

Such projects like those taking place in Austria, Italy and Serbia can serve as best-practice examples for other regions or countries with similar structures and problems. Within BeNatur best-practice projects were collected and exchanged between partners as well as know-how about management of protected areas transferred. A good example for management is the Danube Delta Biosphere Reserve, which is classified as World Natural Heritage.

10. Management for Danube Delta biosphere reserve

Danube Delta Biosphere Reserve Authority (DDBRA) is a public institution under the coordination of the Ministry of Environment and Forests, responsible for the management of reserve, including the conservation and protection of the existing natural heritage; promote and support the sustainable use of the natural resources; provision of support, based on the results of research, for management, education, training and services.

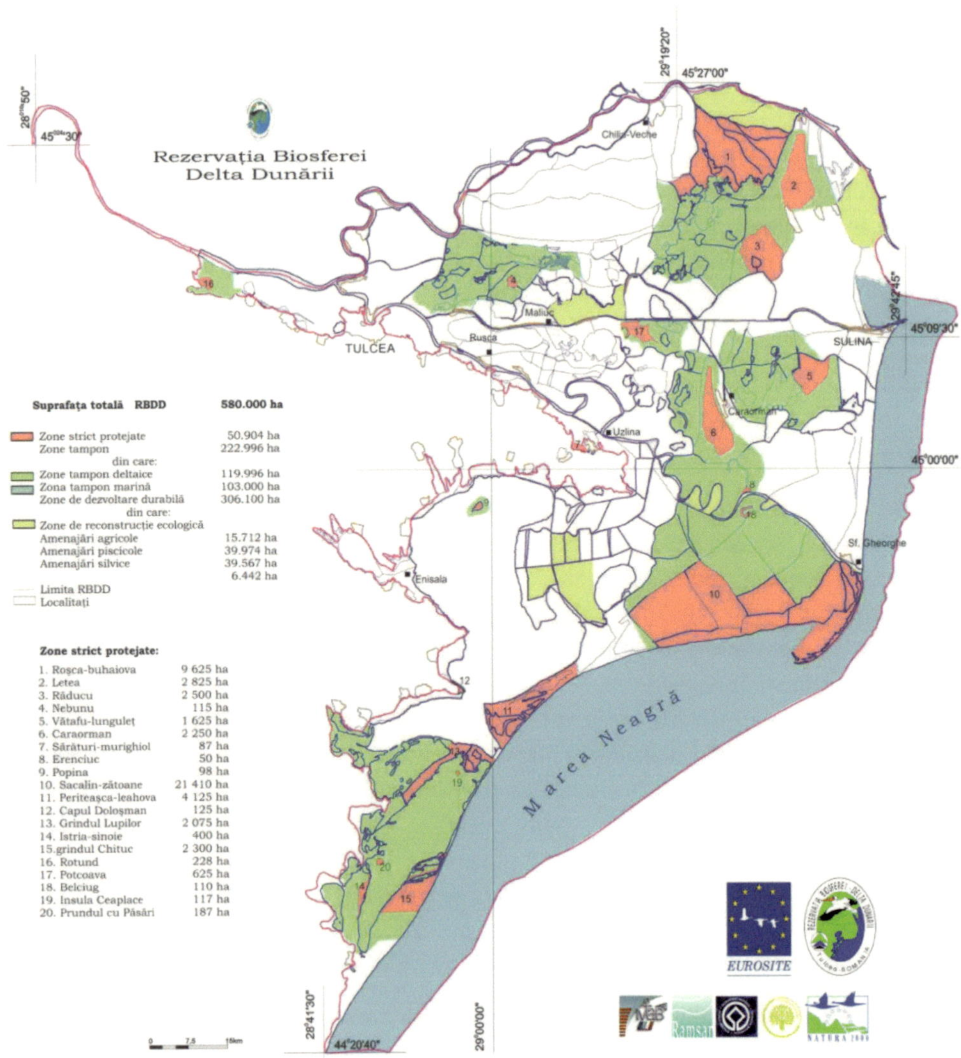

Figure 7. Map of the Danube Delta Biosphere Reserve

Danube Delta was declared a biosphere reserve in 1990 by the Government of Romania and by the Romanian Parliament through the special Law 82/1993 on Danube Delta Biosphere Reserve (DDBR) establishment modified and completed and its international importance is confirmed by its present status.

10.1. General presentation of the area

The surface is about 580,800 ha. By its dimensions, the Danube Delta Biosphere Reserve is the third area in the world in terms of biological diversity with over 5,400 species of plants and animals. It is the most important wetland area in the South-Eastern Europe, having an important contribution to the regional and global water cycle and a unique place where the people are living in isolated settlements spread in the area, close dependent on the natural resources. Danube Delta Biosphere Reserve is a large refuge for migratory birds as a nesting, wintering or resting place, in their way from the Northern Europe to Africa: most of the European population of Common Pelican (Pelecanus onocrotalus) and Dalmatian Pelican (Pelecanus crispus), 60% from world population of Pygmy Cormorant (Phalacrocorax pygmaeus), 50% from world population of Red-breasted Goose (Branta ruficollis), during winter season.

Components: Danube Delta (4,178 km^2 Romania 82%, Ukraine 18%), Razim-Sinoie complex lagoon, Coastal Black Sea up to 20 meters isobaths, Isaccea - Tulcea sector of Danube floodplain, Danube River sector between Cotul Pisicii and Tulcea, Saraturi-Murighiol

Functional areas: Strictly protected areas: 50,904 ha, Buffer zones: 222,996 ha (marine buffer zone: 103,000 ha), Economic zones: 306,100 ha (ecological restoration: 15,712 ha, agricultural polders: 39,974 ha, fish ponds: 39,567 ha, artificial forests: 6,442 ha).

10.2. The international recognition

The universal value of the reserve was recognized by the Man and Biosphere (MAB) Programme of UNESCO in 1990, through its inclusion in the international network of biosphere reserves. DDBR is listed in 1991 both as a wetland of international importance especially as waterfowl habitat under the Ramsar Convention (1991) and as a world natural heritage under the World Cultural and Natural Heritage Convention. Furthermore it received the World Wetland Network Blue Globe Award 2010 for best practice in wetland management (Nagoya, Japan) as well as the QualityCoast Gold Award 2011 for excellence in Nature and Identity a great recognition for Danube Delta Biosphe Reserve, a beautiful area rich in nature, biodiversity and cultural heritage (May 2011, Kouklia, Cyprus).

10.3. The European value

The value of its natural heritage and the efficiency of the management plan applied in the Danube Delta Biosphere Reserve (DDBR) was recognised by the Council of Europe in 2000 who awarded it the European Diploma for Protected Areas, distinction renewed in 2005 and 2010. Besides its inclusion, since its creation, in the information and exchange programs of these networks, the DDBR became member of the EUROSITE who awarded in 1995 for restoration works and in 2001 for management and public awareness. Furthermore it became EUROPARC member and Natura 2000 Site (SPA and SCI).

10.4. Management strategies

The strategy is part of the management plan that deals with the problems of preparation, planning and development for integrating the objectives of the biodiversity's conservation with the implementation of the policies regarding socio-economic systems (SES). The Master Plan regards the integration of the actions for each problem identified in a global strategy that ensure the synergic realization of all the actions to achieve the proposed objectives and goals. The Master Plan's measure of success is given by the implementation degree of the proposed actions.

Action plan to achieve the management objectives in Danube Delta Biosphere Reserve:

a. Management of species and habitats protection
b. Integrated Monitoring
c. Natural resources, sustainable use
d. Tourism and leisure
e. Cultural Heritage
f. Community development and involvement of local population in order to increase their life standard
g. Awareness, Information and education public
h. Cross border cooperation, international cooperation and programmes

An efficient management supports the conservation and sustainable management of the natural resources of the Danube Delta, improving socio-economic benefits for the local population in the Danube Delta, and the financial sustainability of DDBRA, the implementation of the activities for conservation, management and monitoring of Natura 2000 sites and last but not least the improvement of the DDBRA institution capacity and adapted integrated management methods.

Strategic Objectives for 2007 to 2015 are the creation of an integrated monitoring system, an improved public infrastructure, the development of alternative economic activities, the conservation of the rural landscape and the promotion of alternative energy and the restoration of the ecosystem. Further important objectives will be the development of trans-boundary cooperation in the Lower Danube Euroregion and an adaptive management for a better biodiversity conservation.

11. Overall conclusion

In some EU countries provisions for protecting Natura 2000 habitats and species have already been successfully implemented in combination with funding programmes.

Agricultural systems with high natural value (HNV) can aid biodiversity and rural areas. About 25 per cent of agriculturally used areas in the EU (about 27 million ha) are potentially suitable for the designation of HNV, especially EU states with small-structured agriculture. Biological diversity is a public good, therefore farmers must receive a sort of basic pay for the conservation of their bio-inventory according to [37].

The expense of effective implementation of Natura 2000 management is estimated at € 5.8 billion annually at the EU level. The investment should also, however, bring in a high yield of about € 200 billion to € 300 billion per year for the ecosystem services.

The objective should be to integrate the management needs for Natura 2014 starting in 2014 into the Fund for Rural and Regional Development and Fisheries. Some projects were already rudimentarily implemented in the partner countries. The funding programme Life+ will also play a decisive role in the implementation of the Natura 2000 objectives.

It is important that also other policy areas from EU include and support appropriate approaches for nature protection. The basic work for the implementation of requirements from FFH and Birds Directive should be created on a superior level. The management of Natura 2000 has to provide homogenous, comprehensible guidelines as well as measureable criteria for the monitoring of the habitats and species. A comprehensive transnational management of Natura 2000 will be implemented through BeNatur project, which can serve also for other regions as best practice approach for the implementation.

Convincing projects must be developed so that the financing is used optimally. Essentially is the creation of synergies of Natura 2000 with the other land use areas. In reference to [38], the need of management for the Natura 2000 areas must be integrated into agricultural policy so that the added value also can increase in these areas [38].

According to [38] the new Common Agricultural Policy and the agricultural-environmental provisions must steer in the direction of including concerns of biodiversity so that farmers implement environmentally beneficial provisions on the basis of voluntary contracts and also receive appropriate compensation.

The EU countries and regions are invited to adapt their own national and regional funding and management programmes to these new challenges for the integrative Natura 2000 management.

Author details

Renate Mayer*, Claudia Plank, Bettina Plank and Andreas Bohner
Agricultural Research and Education Centre Raumberg, Gumpenstein, Irdning, Austria

Veronica Sărăţeanu, Ionel Samfira and Alexandru Moisuc
Banat University of Agricultural Sciences and Veterinary Medicine Timişoara, Romania

Hanns Kirchmeir and Tobias Köstl
E.C.O. Institute of Ecology, Klagenfurt, Austria

---

* Corresponding Author

Denise Zak
Vienna University of Technology, Austria

Zoltán Árgay, Henrietta Dósa, Attila Gazda, Bertalan Balczó,
Ditta Greguss, Botond Bakó and András Schmidt
Ministry of Rural Development, Hungary

Péter Szinai and Imre Petróczi
Balaton Uplands National Park Directorate, Hungary

Róbert Benedek Sallai and Zsófia Fábián
Nimfea Natura Conservation Association, Hungary

Daniel Kreiner and Petra Sterl
Nationalpark Gesäuse GmbH, Austria

Massimiliano Costa
Parks Office-Province of Ravenna, Emilia-Romagna, Italy

Radojica Gavrilovic and Danka Randjic
City of Cacak, City Administration for LED (Local Economy Development), Serbia

Viorica Bîscă, Georgeta Ivanov and Fănica Başcău
Danube Delta Biosphere Reserve Authority, Romania

Acknowledgement

The chapter "Management for Danube Delta Biosphere Reserve" was elaborated by the Danube Delta Biosphere Reserve Authority working group responsible for the implementation of the activities of the observer (DDBRA). Elaborators are thankful to all collegues' observation, completions, information and ideas for the content of this text and also to our collegues' support from DDNI.

12. References

[1] Commission to the Council and the European Parliament. Composite Report on the Conservation Status of Habitat Types and Species as required under Article 17 of the Habitats Directive. COM(2009) 358 final. July 13, 2009, Brussels.

[2] Rodrigues ASL., Akçakaya HR., Andelman SJ., Bakarr MI., Boitani L., Brooks TM., Chanson JS., Fishpool LDC., Da Fonseca GAB., Gaston KJ., Hoffmann M., Marquet PA., Pilgrim JD., Pressey RL., Schipper J., Sechrest W., Stuart SN., Underhill LG., Waller RW., Watts MEJ., Yan X. Global Gap Analysis: Priority Regions for Expanding the Global Protected-Area Network. BioScience 2004;54(12) 1092-1100.

[3] Blondel J. Biogeographie - approche ecologique & evolutive. Paris: Masson; 1995.

[4] Bănărescu P., Boşcai N. Biogeographie. Bucureşti: Editura Ştiinţifică; 1973.

[5] Harrison I., Laverty M., Sterling E. Alpha, Beta, and Gamma Diversity; Connexions; 2004. Available from http://cnx.org/content/m12147/1.2/ (accessed 14 May 2012).

[6] Hunter Jr ML. Fundamentals of Conservation Biology. Maine USA: Blackwell Publishers; 2002.

[7] Whittaker RH. Evolution and measurement of species diversity. Taxon 1972;21(2) 213-251.

[8] Jost L. Partitioning diversity into independent alpha and beta component. Ecology 2007;88(10) 2427-2439.

[9] Scott JM., Davis F., Csuti B., Noss R., Butterfield B., Groves C., Anderson H., Caicco S., D'Erchia F., Edwards TC Jr., Ulliman J., Wright RG. Gap Analysis: A Geographic Approach to Protection of Biological Diversity. Wildlife Monographs 1993;123 3-41.

[10] Jennings MD. Gap analysis: Concepts, methods, and recent results. Landscape Ecology 2000;15(1) 5-20.

[11] Burley FW. Monitoring biological diversity for setting priorities in conservation. In: Wilson EO. (ed.) Biodiversity. Washington DC: National Academy Press; 1988. p227-230.

[12] Kirchmeir H., Köstl T., Zak D., Getzner M. BE-Natur: BEtter management and implementation of NATURa 2000 sites. WP3: Transnational joint strategy and tools for the better management and implementation of Natura 2000 sites. Individuation of gaps in the management and implementation of Natura 2000 sites (gap analysis). Final research report, Vienna; 2012.

[13] European Environmental Bureau (EEB). Where there is a will there is a way: Snapshot report of Natura 2000 management. Brussels: 2011. 23p.

[14] WWF Natura 2000 in the new EU Member States – Status report and list of sites for selected habitats and species. Brussels: 2004.

[15] Maiorno L., Falcucci A., Garton EO., Boitani L. Contribution of the Natura 2000 network to biodiversity conservation in Italy. Conservation Biology 2007;21(6) 1433-1444.

[16] Mertens, C. (2009) Agency of NGOs in the implementation of Natura 2000 in Hungary. Presented at the Amsterdam Conference on the Human Dimensions of Global Environmental Change 2-4 December 2009 Amsterdam, The Netherlands.

[17] Apostolopoulou E., Pantis JD. Conceptual gaps in the national strategy for the implementation of the European Natura 2000 conservation policy in Greece. Biological Conservation 2009;142(1) 221-237.

[18] Dimitrakopoulos PG., Memtsas D., Troumbis AY. Questioning the effectiveness of the Natura 2000 Special Areas of Conservation strategy: the case of Crete. Global Ecology and Biogeography 2004;13(3) 199-207.

[19] Geitzenauer M. Eine europäische Naturschutzpolitik als Ländersache: Die Umsetzung von Natura 2000 in Österreich. Tag der Politikwissenschaft, DEC 2; Salzburg, Austria; 2011.

[20] Austrian Court of Audit. Implementation of the Natura 2000 Network in Austria: audit report, April and May, 2007, Vienna; 2008.

[21] Bogner D., Golob B. Landwirtschaft in Österreichs Natura 2000 Gebieten - Agriculture in Austrian Natura 2000 Sites. In: Darnhofer I. (ed.) Jahrbuch der Österreichischen Gesellschaft für Agrarökonomie - Band 10. Wien: Facultas Verlag; 2005. p127-135.

[22] Sonderrichtlinie des Bundesministers für Land- und Forstwirtschaft, Umwelt und Wasserwirtschaft zur Umsetzung von Maßnahmen im Rahmen des Österreichischen Programms für die Entwicklung des ländlichen Raums 2007–2013 - „sonstige Maßnahmen" BMLFUW -LE.1.1.22/0012-II/6/2007

[23] Gerecke R., Haseke H., Klauber J., Maringer A. Quellen - Schriften des Nationalparks Gesäuse - Band 7. Weng im Gesäuse; 2012.

[24] Haseke H. Managementplan Revitalisierungsprojekt Johnsbach-Zwischenmäuer 2006-2008. LIFE Gesäuse - Naturschutzstrategien für Wald und Wildfluss im Gesäuse. Weng im Gesäuse; 2006.

[25] Holzinger A., Haseke H., Kreiner D., Zechner L. Managementplan Wald. LIFE Gesäuse - Naturschutzstrategien für Wald und Wildfluss im Gesäuse. Weng im Gesäuse; 2009.

[26] Egger G., Kreiner D. Managementplan Almen. LIFE Gesäuse - Naturschutzstrategien für Wald und Wildfluss im Gesäuse. Weng im Gesäuse; 2009.

[27] Zechner L. A5 Managementplan Besucherlenkung. LIFE Gesäuse - Naturschutzstrategien für Wald und Wildfluss im Gesäuse LIFE05 NAT/A/000078. Weng im Gesause; 2009.

[28] Haseke H., Remschak C. Managementplan Neobiota. LIFE Gesäuse - Naturschutzstrategien für Wald und Wildfluss im Gesäuse. Weng im Gesause; 2010.

[29] Hohensinner S., Muhar S., Jungwirth M., Pohl G., Stelzhammer M. Leitlinie Enns - Konzept für die Entwicklung des Fluss-Auen-Sytems Steirische Enns (Mandling-Hieflau). Wien; 2008.

[30] Haseke H., Kreiner D. LIFE Gesäuse - Final Report - Naturschutzstrategien für Wald und Wildfluss im Gesäuse. Weng im Gesäuse; 2011.

[31] Kreiner D., Maringer A., Zechner L. Econnect - Improving Connectivity in the Alps. Implementation in the pilot region Northern Limestone Alps. Eco.mont 2012;4(1) 37 -42.

[32] Corticelli S. Carta della Vegetazione Parco regionale del Delta del Po - Stazione Pineta di Classe e Salina di Cervia. Servizio Cartografico e Geologico della Regione Emilia-Romagna; 1999.

[33] Escofet A., Espejel I. Conservation and management-oriented ecological research in the coastal zone of Baja California, Mexico. Journal of Coastal Conservation 1999; 5: 43-50.

[34] Schulz R., Stock M. Kentish plovers and tourists: competitors on sandy coasts. Wader Study Group Bull 1993; 68: 83-91.

[35] Brandolini A. Catalogo della mia collezione di uccelli del Ravennate. Stab. Grafico Fratelli Lega, Faenza; 1961.

[36] Costa M., Ceccarelli P.P., Gellini S., Casini L. & Volponi S. Atlante degli uccelli nidificanti nel Parco del Delta del Po Emilia-Romagna (2004-2006). Consorzio di gestione del Parco regionale del Delta del Po; 2009.

[37] Buckwell A. Landwirte sind die Hüter des Ländlichen Raumes. Umwelt für Europäer 2010; ISSN 1563-4175 12p.

[38] Potočnik J. Sicherung der zukünftigen Finanzierung des Natura 2000 Netzwerkes. Natura 2000 Newsletter Natur und Biodiversität der Europäischen Kommission 2012;321(1); ISSN 1026-6178 2p.

Chapter 9

Effectiveness of Nature Conservation – A Case of Natura 2000 Sites in Poland

Małgorzata Grodzińska-Jurczak, Marianna Strzelecka,
Sristi Kamal and Justyna Gutowska

Additional information is available at the end of the chapter

http://dx.doi.org/10.5772/50664

1. Introduction

Accession to the European Union (EU) provided the Member States with new and extensive opportunities for policy development as well as changes in the management of their national, regional and local economies. The EU Member States had to implement standards of the European Union law, which included a broad spectrum of principles of sustainable development [1]. Specifically with regard to nature conservation, the European policy strengthened the implementation of a rational development strategy by influencing the Member States to adopt international commitments such as the Convention on Biological Diversity, and through the expansion of nature conservation areas. Among the EU directives promoting nature conservation, the most important provisions were the Birds Directive and the Habitats Directive. Implementation of these two directives subsequently gave rise to a new form of nature conservation — the Natura 2000 European Ecological Network.

At the regional level of the EU, the general principles and the implementation of the nature conservation policy are complex and governed in a top-down manner. Such approach is inherently at risk of being introduced locally with a low level of effectiveness and adaptability. Hence, current mechanisms of nature protection (mainly biodiversity) in the EU need to be complemented with effective bottom-up initiatives in addition to new means of top-down approaches. The latter appear to be essential, particularly in the new Member States where nature conservation is still affected by the post-socialistic governance and it operates in a rather ineffective way [2].

Recognizing the importance of and integrating the social dimension with the ecological needs, we observe a slow shift in the nature conservation paradigm toward increasing the participation of local stakeholders for more locally sustainable outcomes [3]. For locally sustainable environmental policy solutions, stakeholders' participation in nature conservation is essential. One of the issues evident from the practice of countries that introduced the new nature conservation policy - the EU-25, seems to be the involvement of the possibly large group of stakeholders at all levels of decision-making (local governments, communities, business, non-governmental organizations etc.), but with special attention to local level processes related to the Natura 2000 Network [4]. Within the sustainable development paradigm, the EU public participation is both means to achieve sustainability and the leading principle of rural development.

The concept of nature conservation has changed from strictly traditional, biophysical perspective towards a more innovative approach that integrates the protection of flora and fauna, and habitats with social and economic activity [5-7]. However, natural resource conservation in Poland has been traditionally focused on the preservation of natural environment without deeper consideration of the interests of local stakeholders, who are an important component of those environments. Development of policies concerned with environmental protection adopted the top-down model of decision-making, which implies that stakeholders such as local authorities, environmental groups operating in rural localities as well as owners of the private land under protection have little impact on land designation and management. The authors seek to develop a report based on the available studies and the authors' experience with the European Ecological Network - Natura 2000 that builds the discussion framework to examine problems emerging due to the designation of protected areas as well as implementation and management of the Natura 2000 in Poland.

2. The ecological network natura 2000 in the European Union

The Ecological Network Natura 2000 is the most recent form of the nature conservation strategy implemented in the European Union Member States. It differs considerably from the previous traditional protection system in that it aims at halting the biodiversity loss and maintaining or reconstructing the favorable nature conservation status by protecting natural habitat types, besides protection of floral and faunal species that are unique in the European continent. The popularity of the European Ecological Network after this time period is still debatable [8, 9]. It includes sites designated according to two nature conservation directives of the European Union. Bird Directive (79/409/EGK) accepted in 1979 refers to specific birds' habitat as Special Protection Areas (SPAs), while Habitat Directive (43/92/EGK) from 1992 led to designation of Special Areas for Conservation (SACs). As a form of area-focused environmental protection the Natura 2000 is the first international network at a continent scale that is managed independently at the national level. Currently it comprises over 26,106 sites and covers 17.5% of the territory of EU Member States [10].

The beginnings of the Natura 2000 reflected the changing approach to the structure and functioning of especially valuable natural landscapes in European Membership Countries. At first the process of designating Natura 2000 sites was slow due to the lack of agreement on the methodology to evaluate site proposals. Many EU Member States were subjected to legal proceedings for their slow designation rates [11]. Scientific criterion for the selection of sites for Natura 2000 was agreed as the only criterion for choosing the Natura 2000 sites, and these criteria are listed in Annex III of the Habitats Directive. Moreover, even though sites for the Natura 2000 Network were selected on the basis of the same designation criteria, the share of land selected for protection within the Nature 2000 Program significantly varies among the EU Member States. For example, it includes 7.1% of the country's area in the UK, 12.8% in Germany, 20.9% in Portugal to as much as 34.9% in Bulgaria and 35.5% in Slovenia [12].

The selection process reflects solely the ecological emphasis on maintenance of given species or habitat (these are for example: "*size and density of the population of the species present on the site in relation to the populations present within national territory*" and "*degree of representativeness of the natural habitat*"). Despite some consultations with the local governments and citizens about designating areas under the Habitat Directive, Natura 2000 has been viewed a top-down policy that is not considerate of the local communities' needs. Such a situation has led to two types of conflicts: a) vertical conflicts (disagreements between national and local or regional authorities) and b) horizontal conflicts between stakeholders from public and private sectors. In the vertical conflict, local authorities disagree with the methodology adopted to designate sites for the Natura 2000 Network, while horizontal conflicts of interests occur between public administrations such as General and Regional Directorates for Environmental Protection (GDEP; RDEP), which are responsible for implementation of the national law, together with local governments that conform to RDEP's instructions and entrepreneurs, land owners or other private sector stakeholders.

Several examples from across EU demonstrate man-nature conflicts during the planning and implementation of Natura 2000. Germany for instance, has had strong local opposition to the designation of Natura 2000 sites. Farmers depending on established systems for agri-environmental schemes feared that these would no longer apply or become more difficult to access [13]. This fear resulted from little or no communication, due to the Länder (provinces) governments having underestimated the need for adequate stakeholder information and the associated administrative commitment [14]. Similarly, in France, the implementation of the network was questioned by a number of stakeholder groups (including important representatives from the agricultural, forestry, game and fish-breeding sectors) and ultimately caused the national suspension of the directive. In 1996 protesting groups drafted a declaration taking up the claims. While reasserting the fact that they were not opposed to the principle of conservation, they objected to the methods used to compile the list of sites and the extent of surface areas involved. They demanded the surface areas of the Natura 2000 sites to be reduced and financial resources to be allocated so as to compensate for the loss of earnings due to the new management measures [15]. Other examples of disagreements due to Natura 2000 include Finland, where the network caused major conflicts between landowners mainly lumberjacks and environmental authorities, and ultimately affected countrywide attitudes towards biodiversity conservation [14-16].

Apart from the conflicts related to Natura 2000 within Europe, the El Teide Declaration from 2002 highlighted the key factors crucial for successful implementation of the program, which included: *"the success of Natura 2000 will require the support of European citizens, especially of local people and landowners, and their participation in the decisions on the implementation of the conservation and management of the areas involved"*. It also indicated that: *"many of our valuable Habitats are the result of traditional land use and their conservation relies on traditional practices and skills"*. Current Member States and then the Candidates to the EU (Bulgaria, Cyprus, Czech Republic, Estonia, Hungary, Latvia, Lithuania, Malta, Poland, Romania, Slovakia, Slovenia and Turkey) that signed the document committed to *"promote awareness and understanding of Natura 2000"* as well as: *"promote the development of partnerships involving the broad range of stakeholders in the conservation and management of Natura 2000 sites"*. Whereas in the "old" 15 EU Member States the conflicts between stakeholders in Natura 2000 have been mitigated, countries such as Poland continue to struggle with the program's arrangements, while looking for the most suitable solutions.

3. Natura 2000 network in Poland – A success story?

The problem of nature conservation in Poland is not new, but following the EU accession, the public participatory approach to biodiversity management has become a legislative requirement (Environmental Law - *article 158;* Act 2000 on Access to Information on the Environment and Its Protection and on Environmental Impact Assessment – *article 4, 13,* Law on Public Information). In this light, Natura 2000 Network has become a controversial issue in a number of rural areas. This situation usually happens when the principles of implementation of the European Ecological Network are considerably different from traditional forms of environmental protection [17,18]. Currently Poland, similarly to other Central Eastern European countries, is challenged by rapid social and institutional change, conflicts between traditions of centralized decision-making and new public values and concerns [2,19].

From the very beginning, Natura 2000 Network in Poland caused problems with its' acceptance mainly due to the significant difference from a considerably well-established conservation system over the country and due to the ownership structure of the land covered by the new protected areas. In fact, only biological scientists placed much hope in the program, expecting it to make the protection of native species and habitats more effective on the strength of national legislation, if they are also protected outside Poland. Others, such as local governments of municipalities with areas covered by Natura 2000 Network and local stakeholders perceived it as a threat to local and regional socio-economic development. From their perspective, the program would introduce restrictions on developments in municipalities by creating barriers to usage of one's land and curtailing production and investments. A general negative attitude to the program has not changed much till now [17,18].

The initial step to implement Natura 2000 Ecological Network already begun during late 90s, and the first stage was the preliminary analysis of habitats and species that would require protection. Poland was also negotiating for filling the gap in the EU policy about protected habitats and species that do not occur in any of the "old" Member States of EU and that had not been included in the contemporary nature protection systems. After the initial site identification process, the first phase of the Natura 2000 implementation in Poland focussed on designation and monitoring of the Natura 2000 sites. The boundaries of the sites included in Natura 2000 Network were primarily defined based on biological criteria, without seeking input from local societies or local governments [20-22]. The process was completed mainly by representatives of a few national research institutions and ecological non-governmental organizations (ECO-NGOs). In principle, protection objectives and methods should have been to some extent adapted to local social, economic and cultural conditions [23], however, the process did not consider the existing physical development plans. Moreover, the program's implementation plan did not take into account the possibility of social conflicts and consequently it did not provide for means of prevention [24]. Conflicts started to develop during designation of the sites boundaries and continued during the creation of individual areas' protection and management plans [25].

The Natura 2000 Network implementation procedures and timeline have been in force in Poland since the country's 2004 accession to EU, just as they had been in force in other EU Member States. The Polish Ministry of the Environment requested the local authorities to evaluate the boundaries of Natura 2000 sites within their territories. Majority of boroughs expressed a negative opinion of the designation process and its outcomes. They believed that the sites' designation methodology applied rather old-fashioned and un-professional consultation strategies in the form of one-way written opinion letters delivered to the Ministry by the municipal authorities. Neither did direct consultation with municipal governments take place, nor were they provided with any response regarding proposed changes [26]. Disregarding objections, in May 2004 the Polish national government forwarded the proposal of Natura 2000 Network to the European Commission. The updated version of the document led to strong opposition from experts involved in the creation of its first version. In response, several ECO-NGOs (Klub Przyrodników; PTOP Salamandra), prepared another proposal popularly referred to as the "Shadow List 2010" during the Bilateral Bio-geography Seminar in Warsaw, and independently sent it to the European Committee and the institution responsible for Natura 2000 operation in Poland: General Directorate for Environmental Protection - GDEP. Their list comprised of additional 33 sites Natura 2000 and modified boundaries of 22 areas. It consisted of land that needed to be added to the Natura 2000 Ecological Network according to conclusions from the seminar and findings from a number of projects funded by EU. The European Committee acknowledged both lists and a combination of both proposals (a preliminary and Shadow) was approved. The proposal of the Shadow List provoked further tensions between the ministerial authorities and the experts - mainly NGOs representatives, which one more time delayed the designation of the boundaries of protected areas.

The list was finally delivered to the EU Commission for approval in late 2009, after the Commission issued warning to the Polish government over its insufficient progress in implementation of Natura 2000 as well as a notice about violation of Birds Directive due to insufficient designation of Special Protection Areas (SPAs) [27]. Faced with a lack of response from the Polish authorities to these warnings, the EU Commission went to the

Court of Justice of the European Union in Luxemburg. Determination of the European Committee intensified work on completion of the list of protected areas during the following years and Poland completed the list of Special Areas of Conservation (SACs) in 2010. Currently Natura 2000 Ecological Network covers 19.8% of the country. It includes 823 SACs and 144 SPAs for special birds' protection. The Natura 2000 forced some administrative changes in General Directorate for Environmental Protection. However, these changes in the management structure of the GDEP has had limited impact on management practices in Poland. Despite the fact that new governmental bodies are now responsible for the management of Natura 2000 areas (the General Directorate for Environmental Protection and its representatives in each province: Regional Directorates for Environmental Protection, directors of national parks, directors of marine administration as well as the Forest District), there is a gap in innovative strategies to decrease the friction between local institutions and agencies in implementation, management and monitoring of Natura 2000 sites. Although the agencies play an important role in the management process, the management efforts are still ineffective and it remains unclear what can be done to improve it. The authors seek to explain the most prevailing causes of the controversial nature of Natura 2000.

4. Designation of Natura 2000 network in Poland - Conflicts and misunderstandings

The tasks of the Natura 2000 Network are implemented jointly with provincial and local governments. Local authorities (Regional Directorates for Environmental Protection) are responsible for creation and administration of protected sites at the provincial level as well as monitoring and protection of floral and faunal species. So far, in Poland, protected sites have been established and supervised independently of local authorities. Although, the recently gained experiences have revealed many advantages of delegating some environmental protection responsibilities to local governments, officials have insufficient skills and limited budgets [28]. Natura 2000 Network was designated in 966 boroughs (out of total 2479 municipalities in Poland), and in some cases Natura 2000 sites cover surface of an entire borough. Thus, the engagement of different groups of stakeholders in nature conservation management should be one of the national priorities. So far there have been only a few promising initiatives from organizations such as Sendzimir Foundation [29] or ICLEI – Local Governments for Sustainability (http://www.iclei-europe.org) that supported the implementation of sustainable development principles in Poland by adapting a bottom-up approach or community-driven development. Unfortunately, there aren't any bottom-up initiatives addressed in communities within the Natura 2000 sites as well as the areas that border with Natura 2000 Network.

Taking under consideration the limited time to oppose designated boundaries for land protection, it is reasonable to state that the designation of protected sites occurred without prior consideration of the local views and stakeholders' needs. In fact, no public consultation or other form of participation in decision-making took place [17]. Only in few regions (mainly southern parts of the country), selection of Natura 2000 sites was conducted via opened public consultation aimed at incorporating a broad spectrum of actors (local stakeholders, private landowners, NGOs). In these cases, borders of sites proposed by the nature experts were negotiated and finally changed according to the locals' suggestions [30].

Currently, Poland is in the second phase of the Natura 2000 Network - characterized by development of management plans for designated protected sites. These plans seem to be especially controversial as they have direct effect on boroughs' local economies. Residents as well as other stakeholders were forced to maintain the habitats requiring active protection in their proper condition due to the Natura 2000 Network requirements (e.g. intensive or extensive agriculture activities). However, due to a top-down approach their interests in collaborating on development of area management plans had been neglected. Active protection on private lands is impossible without prior agreement and support of landowners [26,30,31]. As much as public consultations are anticipated before development of protection plans in Poland, it is still confusing which tools will be used. Also, the effectiveness of such consultations as a form of stakeholders' participation in environmental decision-making is questionable.

5. Compensating boroughs' economic loss due to Natura 2000 network

Long-term sustainable development has not been and will never be easy for peripheral areas due to low quality infrastructure, low levels of entrepreneurship, as well as residents' mobilization and no motivation for joint actions [32]. To date only few studies have shown evidence of Natura 2000 having negative impacts on boroughs financial condition. However, it is clear by now that formal limitations due to the Ecological Network impact rural economies, including a borough's income. More and more local leaders demand reimbursement for costs of protection of habitats and species [33].

Local governments associated in Rural Communes of the Republic of Poland [42], proposed the introduction of ecological fiscal transfer. The essence of this financial tool is the redistribution of funding from national to local authorities to compensate for income loss for some local governments due to the large share of protected land. Such a financial tool has been successfully introduced in Portugal, Germany and Brazil [34-36]. Other countries that have not introduced reimbursement programs, attempt to deal with ownership conflicts within the protected areas (the most commonly in newly established or enlarged national parks and Natura 2000 areas) by employing tools such as negotiations, mediations or financial compensation. These solutions are most common in Great Britain, France and Finland [8,15,37,38,39]. Yet, none of the European countries has created a solution that would satisfy all stakeholders. Ecological fiscal transfer in Poland would let the municipalities manage their Natura 2000 sites in an effective way [34,40-41]. Also, similar initiatives could be the first step in undertaking a participatory approach to biodiversity conservation in Poland, while responding to much needed change in environmental management.

The Rural Communes of the Republic of Poland whose municipalities are included into Natura 2000 Network raised an official objection, protesting against short notice to formulate opinions, use of pure scientific criteria (marginalizing economic and social aspects) while selecting the protected areas, and the system of financing Natura 2000 (no economic schemes to encourage local authorities and private owners to support nature protection or to compensate lost profits) [26,30,42]. Due to ineffective top-down distribution of funding resources for completion of the Natura 2000 tasks no resources reached the localities [40]. The opinion of the Supreme Control Chamber is that the funds management has been insufficient, and the current spending on Natura 2000 is underestimated as the expenses of a variety of institutions (e.g. local governments, NGOs, national parks and National Fund of Environment Protection and Water Management) were excluded [41].

Dissatisfaction with the implementation of Natura 2000 Network led to consolidation of local governments and taking an initiative on the above mentioned ecological fiscal transfer proposal. In Poland, presence of protected areas decreases gross boroughs income and stakeholders' annual income from the protected areas compared to the Special Economic Zones (SEZ) municipalities. Designation of the protected areas has not been backed by any appropriate national financial policy while local governments are expected to complete various tasks towards nature conservations on their land while the SEZ boroughs attract potential investors by an economically profitable tax allowance system.

The Council of the Rural Communes of the Republic of Poland representing municipalities situated in the regions with protected areas developed a proposal for a fiscal transfer mechanism - Ecological Subvention Act - a tool of sustainable development policy. The Act proposes ecological subventions – a type of financial compensation for municipalities whose protected parts of the territory are excluded from a business activity. Ecological subventions are to be spent without any limitations on a variety of local governmental needs, and to support a range of local investments. Invested resources would be reimbursed into the national budget in the form of 23% VAT tax and personal tax to subcontractors. Calculation of ecological subventions would be based on algorithms proposed by the Ministry of Finance, *en vertu* on a proposed act. It is assumed that completion of the Ecological Subvention Act will result either in extra expenditures or shifting resources within the current national budget. It does not, however, cause, additional expense to local authorities. A total expenditure of national budget for ecological subvention initiative is approximated to be ca. 200 mln EURO. The ecological subvention proposal was widely consulted with General Directorate for Environmental Protection, members of the Polish Parliament, representatives of national and regional governments and lawyers. This bottom-up initiative was also highly regarded by the President of Republic of Poland [42].

Currently the process of designing compensation for designated sites Nature 2000 in Poland, including the Ecological Subvention Act, is focused on municipalities and local governments. However, another contentious issue in the country with regards to Natura 2000 sites is its occurrence on private lands. While defining the potential of conflict over private land involved in conservation in Poland, the following section looks into the existing instruments being explored to deal with private land conservation in other countries.

6. Conflicts over conservation on private lands

From the global perspective, the success of nature protection and in-situ biodiversity conservation relies heavily on protected areas. Since most of these areas are under government authorities and agencies, it has led to the common belief amongst stakeholders that the responsibility of maintaining the functional elements promoting nature conservation lie on the government. However, protected areas in in-situ conservation are limited by the fact that they occupy only 12.5% of the global land cover [43], often fragmented and isolated from one another, and they support only a fraction of the biodiversity and ecosystem services. In Poland, 32.4% of the total land area is under some form of legal protection for nature conservation. However, the ownership structure diverges from the conventional assumption that protected areas in the country are usually state owned. For example, 15.9% of the national parks' land area is under private ownership [44]. It is expected that significant portion of the Natura 2000 areas lie on private land as well [12,44]. Hence, the final issue linked to the designation of Natura 2000 areas is the protest of private land owners against rules enforced by the Network that affect their economic wellbeing.

Typically, any planning strategy focuses only on the ecological system and not the broader socio-ecological systems, which is where conservation in reality occurs. Hence, without involving all stakeholders in the decision making process, what needs to be conserved against what can be conserved becomes a debatable issue. Private lands with their larger land coverage, have a strong potential in promoting biodiversity conservation and maintaining habitats and their connectivity [45]. They can make substantial contributions to biodiversity conservation needs and therefore, need to be integrated into the conservation strategies [46]. On the other hand, private land could also be a serious threat to biodiversity due to the deforestation and other land-use changes and more actions need to be directed towards encouraging preservation of nature [47]. In these circumstances, integrating stakeholder participation in planning and decision-making becomes crucial for effective conservation actions. This requires looking towards a more comprehensive bioregional model that conserves landscapes, irrespective of the nature of ownership [48]. However, integrating private land into conservation planning and management is complicated by the nature of ownership of the land and the complex social and economic traits that are inter-related with its current use [49-51]. The land use structure in the Natura 2000 areas lying on private land, for instance, are managed by their owners, chiefly farmers [52]. The issue of private land conservation has been explored through a diverse spectrum of mechanisms such as regulatory prohibitions and requirements including use of legal instruments, government acquisition of land or right over resource use leading to relocation and rehabilitation of previous residents (as observed during the establishment of the first national parks in the USA and still practiced in developing countries such as in Central Africa and South Asia), direct incentives for private conservation action [53-56] or educational programs and public consultations.

Globally, conservation on private land has been one of the main reasons for conflict, as it raises the issue of development and property rights versus the restrictive approach to conservation and to address this, both involuntary and voluntary tools have been used. Involuntary actions include relocation of people from private land with conservation value through direct purchase of the land by the government. This is usually accompanied by rehabilitation negotiations between landowners and authorities [55,57]. Another popular tool is regulations or restrictions directed towards landowners on the usage of their land. This top-down approach appears less intrusive, but it is nevertheless an issue of contention over property rights and right of use of the land. Government usually have limited budgets to acquire the land and so they prefer a mixed model of private and public protected areas, where private lands included in protected area are subjected to the same restrictions as public lands [58]. Involuntary acquisition and imposed regulations have been the primary strategies for conservation of nature in Poland. In Natura 2000, the sites have been designated based on their ecological significance and scientific opinion leading to considerable proportion lying on private lands in several EU countries and Poland is no exception [17,54]. Usually, the regulations and restrictions imposed over the public land within the park also become applicable to a large extent on the private land situated within the strict protected area [54,58] and subsequently, it has met with strong resistance from private landowners who see no direct benefit from their land being included in protected areas [17,59].

Voluntary tools include new strategies that provide an incentive to landowners to involve them in the process such ilegally establishing private reserves, use of conservation easements, forest certification for forest products and conservation contracts, to name a few [46,54,56,60-66]. The success of these tools in addressing the conflict of development versus conservation has met with a varying degree of success in different regions. While private reserves, including game ranches, are very popular in Africa and in some Central American countries owing to the presence of mega-fauna [64-65,67], the use of conservation easements on the other hand, has been more popularized in developed nations such as USA, Australia and to some extent in the UK [46,56,63,68]. The use of such tools has not been documented in Poland or other Central and Eastern Europe countries. One reason could be that use of tools such as conservation easements requires financial support from national or regional authorities in order to compensate for the deficit incurred by local administration in the form of lower tax collection due to tax reliefs that these easements typically offer. This would require their respective governments to direct more financial resources towards nature conservation, and often these nations are limited in their budget. Forest certifications as an incentive based tool also has a global appeal with FAO reporting 7% of the world's private forests being certified by 2006. International certification agencies such as The Forest Stewardship Council (FSC) have a presence in several countries, including Poland. However, the cost certification and lack of consumer awareness about certified products have been the primary challenges in promoting this tool more efficiently.

Perhaps the most common conservation tool being used in Europe has been the conservation contracts. These binding voluntary agreements are signed between a landowner and a government agency/authority to conserve the natural features on the land, or encourage activities with a conservation core on private land in return for incentives such as technical help, finances for weeding etc. Besides national level contracts such as Austria's Natural Forests Reserve Program and Sweden's Nature Conservation Agreements, the largest of tools in terms of its scale is the Agri-Environmental Scheme (AES) under the Common Agricultural Plan (CAP) of the EU. It has been implemented in almost all the EU countries. Since the impetus behind this scheme was to promote improved and environmentally sound agricultural practices, AES specifically targets farmers. France, for instance, developed special compensation measures to make conservation on farmlands more attractive after it received strong opposition from farmers over the designation and implementation of Natura 2000 sites, significant proportion of which lay on farmlands. This change in approach towards implementation of Natura 2000 generated more support and acceptance towards the Natura 2000 network from this particular stakeholder group.

The role of agriculture in employment in Poland has shrunk from 26.4% in 1984 to 16.2% in 2005 with a decrease of 5% in agricultural production; however the trend has begun to stabilise and even increase in case of animal husbandry after the shift from centralised economy in the 1980s to the present market-based economy [69]. Contrary to many other centrally planned economies, Poland's farmlands remained mostly under individual ownership leading to more number of small subsistence farmers. The support from CAP in Poland started in 2004 under the Rural Development Plan (RDP), which included aids in inputs and outputs that minimised intensive agricultural practices considered to be harmful to the environment, and instead it encouraged agricultural activities that were believed beneficial to the environment or had a conservation core.

The National Agri-Environmental Programme (NAEP) under the RDP states protection of environment and landscapes, and conservation of biodiversity as two of its main objectives. To achieve these objectives, direct involvement of farmers and increasing their knowledge about the AES and its principles become crucial (OECD, 2008). NAEP has had positive impacts on stabilizing the country's agricultural production along with environmental benefits. In its new phase, it has undergone major changes (2007-2013) to reach a larger community of farmers and target Natura 2000 sites and non-Natura 2000 sites separately with different benefits [70-71].

NAEP faces two major challenges while promoting conservation on farmlands. Firstly, general lack of awareness among farmers on detrimental environmental impacts of agricultural practices (with only 30% of farmers being aware of it) [69] and their subsequent impact on biodiversity has been observed as a hindrance in wider coverage of such a scheme. Without being aware of the effects of their practices, farmers are less likely to modify their existing practices. Secondly, most of Poland's farmers are with small land holdings, which makes it difficult for the AES to reach majority of the farmers in Poland: most compensation programs under the AES was available to only 5% of farmers in its first phase [72] and therefore has not been able to reach a significant proportion of private lands that could have an important role in conservation. This challenge in particular highlights the importance of context specific policies to be able to address the situation on ground.

Different tools to promote private land conservation has met with varying degree of success in different countries or region and this fact only emphasises that these tools are context dependent, including the regional context (the country, political history, economic status) and the type of stakeholders involved. While it is generally accepted that defining areas of conservation priority depends on the level of ecological awareness along with political will, the success of conservation initiatives on such areas is a function of the human and social dimensions, such as stakeholders' willingness and capacity to participate [50,73]. It is therefore imperative to differentiate between areas of conservation priority and that of conservation opportunity. Conservation areas with high ecological value as well as high social value require minimal intervention through external aids or tools; however, areas with high ecological value and low social value will require some incentives to make conservation more attractive and plausible [50,74].

In Poland, especially in the case designation of Natura 2000 sites, biological significance has been the criterion for designating areas for conservation. However, the real potential of what can be conserved remains questionable. With 19.26% of the total Natura 2000 area lying on agricultural land [75], imposing restrictions on land use cannot be the solution. Already, there have been several instances of protest and hostility towards the Natura 2000 network, and this can be attributed to the fact that the process of site designation did not allow for stakeholder participation [17]. Lack of awareness about the Natura 2000 Network adds further to this hostility since landowners now understand the program only as an intrusion into their private space and a violation of their property rights [59]. Although drafting the management plans for these sites through a consultation process is now a legal requirement, thereby allowing for stakeholder participation, successful outcome of such an initiative is often hindered by the fact that the most consultation processes are not handled properly, coupled with the preconceived notion among stakeholders about such protected areas being a hindrance to livelihoods and property rights, which makes the process difficult.

The overall land use structure in Europe has been changing to accommodate for the economic development, and although forest land cover has increased, only 1.6% of the continent's natural forests are protected legally [76- 77]. Poland is no exception to this developmental trend and with its accession to the EU, intensification of certain practices such as those in agriculture is expected. In such a situation, involvement of people in conservation will play an important role in furthering conservation goals [51]. Besides political support at a national level, and financial support at a regional level, it is necessary to find stakeholders supporting long-term sustainable implementation of management plans for protected areas located on private lands.

The authors pose that policy-makers need to identify the factors that increase stakeholders' acceptance of conservation practices on their private land. This will require the research into socio-demographic and economic features of landowners as well as land characteristics (type of land use, type of protected area) [50]. The challenge in private land conservation is to promote conservation values on a land without compromising its capability to meet the requirements of the owners from it. Tools and mechanisms that compensate for the conservation opportunity and that increase social acceptance of the 'protected areas on private land' model are necessary under circumstances where ecologically significant private land in Poland generates direct or indirect economic benefits to its owners.

7. Conclusions

The new environmental policy can only be successful in the Central and Eastern European countries if it is legitimate. Therefore, Poland as well as other post-communist democracies need to re-focus its environmental policy practices toward community empowerment in environmental decision-making which is conceptualized as a process in which community members, who share physical spaces, experiences and concerns, gain influence over conditions that matter to them [78]. Good policy-making requires, among others, up-to-date knowledge or assessment of the "winners" and "losers". Furthermore, its implementation at the local level requires local skills and local resources. The authors propose that legitimate policy must empower communities through participation in environmental management decision-making.

Community used to be defined as a geographic concept or a form of a collective interest revealed in common views on some issues [79]. The authors understand a community as: "the process of interactions through time with direction toward some more or less distinctive outcomes and with constantly changing elements and structure" [80]. By definition, a community is a process in which community participants focus on the betterment of local stakeholders in the context of the Natura 2000 program.

Scholars suggest that direct participation in decision-making is a condition for individual empowerment [81-82] . Others add that non-direct forms of participation in decision-making can also empower stakeholders [82-83]. Local participation has been a concept of increasing importance since the Brundtland Report in 1987 defined it as an indispensable ingredient of sustainable development. Public participation is consistent with the three-dimensional concept of sustainable development as it allows natural capital to be traded off for economic and social capital. It allows residents to observe more closely and evaluate the current governance system in a better way [84-85]. Such distribution of the decision-making power towards local stakeholders integrates democratic elements into sustainable development of the rural post-communist areas in Poland [86].

Participatory decision-making is a key element of the local democratic practice. As much as literature in recent years emphasizes the need for inclusion of stakeholders in decision-making, it also indicates the importance of fundamental arrangements for this community based management and development. Shared control through the inclusion of community members in decision-making is a key element of empowerment [87]. Authorities that attempt to involve community in decision-making in natural resources management must be able and willing to learn from the community members and to apply instruments that empower residents [88-89]. In Poland, the practice of empowering Natura 2000 stakeholders is still in its infancy. Also scholars from social disciplines rarely mention the idea of empowering stakeholders in the context of changing social and political environment of rural Poland.

Discussed difficulties that the majority of the EU Members have experienced in regards to Natura 2000 Network implicate the more global issues associated with the implementation of sustainable development principles and the practice of empowering stakeholders. The core problem of the current approach to Natura 2000 Network as well as other initiatives toward more sustainable Europe is the decentralization of responsibilities for protection of local nature and simultaneous top-down environmental decision-making that facilitates policy creation at the national-level. Hence, EU must focus on legitimate environmental protection policies by distributing rights over environmental decisions to local authorities and other local stakeholders. Also, the efficiency of existing decision-making tools to mitigate and prevent current or future conflicts regarding Natura 2000 Network needs to be re-examined in the context of transitioning economies of the Central and Eastern European Members of the EU such as Poland. The authors propose empowering stakeholders for Natura 2000 through participation in decision-making processes as a locally implemented solution to this global problem.

In addition to increasing legitimization new environmental policies that follow Natura 2000 itself, public participation leads to the development of multilevel governance in the broader and more interdisciplinary context, the introduction of new institutional structures and financial resources to the civil society [90-91]. The non-homogenous character of a community is the main identified barrier to its successful participation in decision-making [92-93]. To date debates, information sharing and creating space for public opinion are the main instruments of participatory approaches [94]. Collaboration and dialogue with governmental representatives create conditions for equity and thereby space for community feedback and community input in decision making that flows upward toward officials [84,85]. Jointly derived decisions contribute to trust building within community [95-96].

Solving problems at the central level proved so far ineffective, and currently documents such as Strategy for Sustainable Development of Poland till 2025 more explicitly articulate that local leadership institutions need to engage stakeholders in the development in order to achieve local sustainability. The Habitat Directive and the Convention of Aarhus [97-98] notes that public participation should manifest itself in society's access to information about the natural environment and its involvement in successive stages of the implementation of protective measures: from planning to making decisions in management. Moreover, bottom-up approaches to biodiversity management will increase stakeholders' perceived control over the local natural environment and increase felt responsibility for its quality.

Author details

Małgorzata Grodzińska-Jurczak, Sristi Kamal and Justyna Gutowska
Institute of Environmental Sciences, Jagiellonian University, Kraków, Poland

Marianna Strzelecka
College of Merchandising, Hospitality and Tourism, University of North Texas, Denton, TX, USA

Acknowledgement

The following manuscript was developed as part of the research project *"Information, education and communication for the natural environment"* sponsored by the Jagiellonian University (grant no. WRBW/DS/INoŚ/760).

8. References

[1] Larobina MD. A report on Poland and European union accession. Multinational Business Review (ISSN: 0-333-71654-X) 2001;9(2) 8-16.

[2] Kluvánková-Oravská T, Chobotová V, Banaszek I. From Government to Governance for Biodiversity: The Perspective of Central and Eastern European Transition Countries. Environmental Planning and Governance (ISSN: 1756-9338) 2009;19 186-196.

[3] Mannigel E. Integrating Parks and People: How Does Participation Work in Protected Area Management? Society & Natural Resources (ISSN: 0894-1920) 2008;21(6) 498-511.

[4] Paavola J, Kluvánková-Oravská T, Gouldson A. Institutions, ecosystems and the interplay of actors, scales, frameworks and regimes. Environmental Policy and Governance (ISSN: 1756-9338) 2009;19 141-147.

[5] Harwood J. Risk assessment and decision analysis in conservation. Biological Conservation (ISSN: 0006-3207) 2000;95 219-226.

[6] Mascia MB. Conservation and the Social Science, Editorial. Conservation Biology (ISSN: 0888-8892) 2003;17(3) 649-650.

[7] Ostrom E. Governing the commons: The evolution of institutions for collective action. Cambridge: Cambridge University Press (ISSN: 00029092); 1990.

[8] Hiedenpää J. European-wide conservation versus local well-being: the reception of the Natura 2000 Reserve Network in Kavia, SW-Finland. Landscape and Urban Planning (ISSN: 0169-2046) 2002;61 113-123.

[9] Weber N, Christophersen T. The influence of non-governmental organisations on the creation of Natura 2000 during the European Policy process. Forest Policy and Economics (ISSN: 1389-9341) 2002;4(1) 1-12.

[10] European Commission. Natura 2000 Barometer-update August 2011, Natura 2000 Newsletter (ISSN: 1026-6151) 2012;31 8-9. http://ec.europa.eu/environment/nature/info/pubs/docs/nat2000newsl/nat31_en.pdf (accessed 15 May 2012).

[11] European Commission. Nature and Biodiversity Cases - Ruling of the European Court of Justice. Luxembourg: Office for Official Publications of the European Communities (ISBN: 92-79-02561-9) 2006. http://ec.europa.eu/environment/nature/info/pubs/docs/others/ecj_rulings_en.pdf

[12] Generalna Dyrekcja Ochrony Środowiska (General Directorate for Environmental Protection) 2010. Online press release. http://www.gdos.gov.pl/Articles/view/1910/Historia_powstania

[13] Stoll-Kleemann S. Opposition to the designation of protected areas in Germany. Journal of Environmental Planning and Management (ISSN: 1360-0559) 2001;44 111-130.

[14] Henle K, Didier AD, Clitherow J, Cobb P, Firbank L, Kull T, McCracken D, Moritz RFA, Niemela J, Rebane M, Wascher D, Watt A, Young J. Identifying and managing the

conflicts between agriculture and biodiversity conservation in Europe-A review. Agriculture, Ecosystems and Environment (ISSN: 0167-8809) 2008;124 60–71.

[15] Alphandéry P, Fortier A. Can a Territorial Policy be Based on Science Alone? The System for Creating the Natura 2000 Network in France. Sociologia Ruralis (ISSN:1467-9523) 2001;41(3) 311-328.

[16] Nieminen M. BiofACT - Finnish report, University of Jyvaskyla, Department of Social Sciences and Philosophy, Jyvaskyla 2004. http:// www.ecnc.nl/file_handler /documents/original/download (accessed 20 September 2006).

[17] Grodzińska-Jurczak M, Cent J. Udział społeczny szansą dla realizacji programu Natura 2000 w Polsce. Public participatory approach - a Chance for Natura 2000 implementation in Poland. Chrońmy Przyrodę Ojczystą (ISSN: 0009-6172) 2010;66(5) 341-352.

[18] Makomaska-Juchiewicz M. Sieć obszarów Natura 2000 w Polsce. In: Gregorczyk M. (ed.) Integralna Ochrona Przyrody (ISBN: 9788391891490). Kraków: Instytut Ochrony Przyrody PAN; 2007. p165-176.

[19] Lawrence A. Experiences with participatory conservation in post-socialist Europe. International Journal of Biodiversity Science and Management (ISSN: 1745-1590) 2008;4 179-186.

[20] Bell S, Marzano M, Cent J, Kobierska H, Podjed D, Vandzinskaite D, Reinert H, Armaitiene A, Grodzinska-Jurczak M, Mursic R. What Counts? Volunteers and their organisations in the recording and monitoring of biodiversity. Biodiversity and Conservation (ISSN: 1572-9710) 2008;17(14) 3443-3454.

[21] Cent J, Kobierska H, Grodzińska-Jurczak M, Bell S. Who is responsible for Natura 2000 in Poland? – a potential role of NGOs in establishing the programme. International Journal of Environment and Sustainable Development (ISSN: 1478-7466) 2007;6 422-435.

[22] Jermaczek A., Pawlaczyk P. Natura 2000 – narzędzie ochrony przyrody. Planowanie ochrony obszarów Natura 2000. Warszawa: WWF Polska; 2004.

[23] Dimitrakopoulos PG, Memtsas D, Troumbis AY. Questioning the effectiveness of the Natura 2000 Species Areas of Conservation strategy: the case of Crete. Global Ecology and Biogeography (ISSN: 1466-8238) 2004;13 199-207.

[24] Bernacka A., Jermaczek A., Kierus M., Ruszlewicz A. Uspołecznione planowanie ochrony przyrody na obszarach sieci NATURA 2000. (Social plans of nature protection at NATURA 2000 sites). Świebodzin: Wydawnictwo Klubu Przyrodników (ISSN: 1230-509X); 2004.

[25] Baranowski M. Prace nad siecią NATURA 2000 w Polsce. In: Makomaska-Juchiewicz M., Tworek S. Ekologiczna Sieć Natura 2000. Problem czy szansa? Kraków: Instytut Ochrony Przyrody; 2003.

[26] Grodzińska-Jurczak M, Cent J. Expansion of Nature Conservation Areas: Problems with Natura 2000 Implementation in Poland? Environmental Management (ISNN: 0364-152X) 2011;47 11-27.

[27] Grodzińska-Jurczak M. Rethinking of nature conservation policy in Poland – the need of human dimension approach. Human Dimensions of Wildlife (ISSN: 1533-158X) 2008;13(5) 380-381.

[28] Trzeciak M. Projektowane regulacje prawne zagrożeniem dla ochrony przyrody. (Risk of proposed regulations to the nature protection). Przyroda Polska (ISSN: 0552-430X) 2005;2.

[29] Kronenberg J., Bergier T., editors. Wyzwania zrównoważonego rozwoju w Polsce. Kraków: Fundacja Sendzimira (ISBN: 978-83-62168-00-2); 2010.
[30] Cent J, Grodzińska-Jurczak M, Nowak N. Ocena efektów małopolskiego programu konsultacji społecznych wokół obszarów Natura 2000. Chrońmy Przyrodę Ojczystą (ISSN: 0009-6172); 2010;66(4) 251-260.
[31] Pawlaczyk P., Jermaczek A. Natura 2000 – narzędzie ochrony przyrody. Planowanie ochrony obszarów Natura 2000. Warszawa: WWF Polska; 2004.
[32] Bołtromiuk A. Uwarunkowania zrównoważonego rozwoju gmin objętych siecią Natura 2000 w świetle badań empirycznych. Warszawa: Instytut Rozwoju Wsi i Rolnictwa Polskiej Akademii Nauk (ISSN/ISBN: 83-89 900-41-6); 2011.
[33] Bołtromiuk A., Kłodziński M. Natura 2000 jako czynnik zrównoważonego rozwoju obszarów wiejskich regionu Zielonych Płuc Polski. Warszawa: Instytut Rozwoju Wsi i Rolnictwa Polskiej Akademii Nauk (ISSN/ISBN: 83-89900-40-8); 2011.
[34] Ring I. Compensating Municipalities for Protected Areas Fiscal Transfers for Biodiversity Conservation in Saxony, Germany. GAIA (ISSN: 0940-5550) 2008;17(S1) 143–151.
[35] Ring I, Drechsler M, van Teeffelen AJA, Irawan S, Venter O. Biodiversity conservation and climate mitigation: what role can economic instruments play? Current Opinion in Environmental Sustainability (ISSN: 1877-3435) 2010;2 50-58.
[36] Santos R, Ring I, Antunes P, Clemente P. Fiscal transfers for biodiversity conservation: the Portuguese Local Finances Law. UFZ-Diskussionspapiere 2010 (ISSN: 1436-140X);11.
[37] Björkell S. Resistance to Top-Down Conservation Policy and the Search for New Participatory Models. The Case of Bergö-Malax' Outer Archipelago in Finland. In: Keulartz J., Leistra G. (eds.) Legitimacy in European Nature Conservation Policy. Case Studies in Multilevel Governance (ISSN: 1570-3010). Wageningen, The Netherlands: Springer; 2007. p109-126.
[38] Fisher R.J. Collaborative management of forests for conservation and development. Issues in forest conservation (ISSN: 1436-140X). Gland, Switzerland: IUCN and WWF; 1995.
[39] Kellert SR, Mehta JN, Ebbin SA, Lichtenfeld LL. Community natural resource management: promises, rhetoric and reality. Society and Natural Resources (ISSN: 1521-0723) 2000;13 705-715.
[40] Chmielewski T.J. Nature conservation management: from idea to practical results. Lublin, Łódź, Helsinki, Aarhus: Alternet, PWZN; 2007.
[41] Najwyższa Izba Kontroli. Informacja o wynikach kontroli wdrażania ochrony na obszarach Natura 2000. http://www.nik.gov.pl/kontrole/wyniki-kontroli-nik/kontrole,1664.html
[42] Związek Gmin Wiejskich Rzeczypospolitej Polskiej. Subwencja ekologiczna. http://www.gminyrp.pl/?slang=pl&art=1&m=7&p=2&id=626 (accessed 2 February 2012).
[43] World Database on Protected Areas. Regional and Global Stats for 1990-2010 from the 2011 MDG Analysis. Cambridge, UK: UNEP-WCMC; 2011.
[44] Central Statistical Office. Statistical Information and Elaboration. Warsaw: Environment (ISSN: 0867-3217); 2011.
[45] Smith G, Phillips E, Doret G. Biodiversity Conservation on Private Land: conference proceedings, 14-18 September 42nd ISoCaRP Council Case Studies. Hague, Netherlands: ISoCaRP; 2006.

[46] Clough P. Encouraging private biodiversity: Incentives for biodiversity conservation on private land. Report to the Treasury, New Zealand Institute of Economic Research, Wellington, New Zealand; 2000.

[47] Mieners RE, Parker DP. Legal and economic issues in private land conservation. Natural Resources Journal (ISSN: 1477-8947) 2004;44 353–360.

[48] Figgis P. Conservation on Private Lands: the Australian Experience. Gland, Switzerland and Cambridge, UK: IUCN (ISBN: 2-8317-0779-X); 2004.

[49] Knight RL. Private lands: the neglected geography. Conservation Biology (ISSN: 0888-8892) 1999;13 223–224.

[50] Raymond CM, Brown G. Assessing Conservation Opportunity on Private Land: Socio-economic, behavioural and spatial dimensions. Journal of Environmental Management 2011;92(10) 2513-2523.

[51] Tikka PM, Kauppi P. Introduction to special issue: protecting nature on private lands – from conflict to agreements. Environmental Science and Policy (ISSN: 1462-9011) 2003; 6 193–194.

[52] Makomaska-Juchiewicz M, Tworek S. Ekologiczna sieć Natura 2000. Problem czy szansa? Kraków: Instytut Ochrony Przyrody PAN (ISSN: 0009-6172); 2003.

[53] Kauneckis D, York AM. An Empirical Evaluation of Private Landowner Participation in Voluntary Forest Conservation Programs. Environmental Management (ISSN: 0364-152X) 2009;44 468–484.

[54] Mayer AL, Tikka PM. Biodiversity conservation incentive programs for privately owned forests. Environmental Science and Policy 2006 (ISSN: 1462-9011);9 614-625.

[55] Rangarajan M, Shahabuddin G. Displacement and Relocation from Protected Areas: Towards a Biological Historical Synthesis. Conservation and Society (ISSN: 09724923) 2006;4 359-378.

[56] Rissman AR, Lozier L, Comendant T, Kareiva P, Kiesecker JM, Shaw RM, Merenlender AM. Conservation Easements: Biodiversity Protection and Private Use. Conservation Biology (ISSN: 09724923) 2007;21(3) 709-718.

[57] Karnath KK. Bhadra Wildlife Sanctuary: Addressing Relocation and Livelihood Concern. Economic and Political Weekly (ISSN: 0012-9976) 2005;40(6) 4809-4811.

[58] Environmental Law Institute. Legal Tools and Incentives for Private Lands in Latin America: Building Models for Success. Washington, DC, USA: Environmental Law Institute (ISBN: 1-58576-059-5). http://www.elistore.org/reports_detail.asp?ID510914

[59] Pietrzyk-Kaszyńska A, Grodzińska-Jurczak M, Szymańska M. Factors influencing perception of protected areas – the case of Natura 2000 in Polish Carpathian Communities. Journal for Nature Conservation (ISSN: 1617-1381) 2012 (forthcoming).

[60] Cubbage F, Diaz D, Yapura P, Dube F. Impacts of Forest Management Certification in Argentina and Chile. Forest Policy and Economics (ISSN: 1389-9341) 2009;12(7) 497-504.

[61] Doremus H. A policy portfolio approach to biodiversity protection on private lands. Environmental Science & Policy (ISSN: 1462-9011) 2003;6 217-232.

[62] Forest Stewardship Council. Country specific units: Poland 2011. http://www.fsc.org (accessed 18 January 2012).

[63] Gattuso D.J. Conservation Easements: The Good, The Bad and The Ugly. USA: National Policy Analysis, National Centre for Public Policy Research; 2008.

[64] Langholz J, Lassoie J. Perils and Promise of Privately Owned Protected Areas. Bioscience (ISSN: 0006-3568) 2001;51 1079-1080.

[65] Ramutsindela M. Parks and People in Post-Colonial Societies: experiences in Southern Africa. Springer (ISBN: 1-4020-2843-1); 2004.

[66] Stolton S., Mansourian S., Dudley N. Valuing Protected Areas. Washington DC: The World Bank; 2010. http://www.cropwildrelatives.org/fileadmin/www.cropwildrelatives.org/documents/Valuing%20Protected%20Areas.pdf (accessed 24 December 2011)

[67] Langholz J, Krug W. New Forms of Biodiversity Governance: Non-state actors and the Private Protected Area Action Plan. Journal of International Wildlife Law and Policy (ISSN: 1388-0292) 2005;7 9-29.

[68] Reid CT. The Privatisation of Biodiversity? Possible New Approaches to Nature Conservation Law in the UK. Journal of Environmental Law (ISSN: 0952-8873) 2011;23(2) 203-231.

[69] OECD. Environmental Performance of Agriculture in OECD countries since 1990. Paris, France: OECD; 2008.

[70] Goliński P., Golińska B. Agri-environmental funding schemes – a tool for supporting the conservation of semi-natural grassland in Poland; 2011. http://www.egf2011.at/files/pubs/592_golinski.pdf (accessed: 6 February 2012).

[71] Liro A. Conditions of natural environment in rural areas. In: Wilkin J., Nurzyńska I. (eds.) Rural Poland 2010, Rural Development Report. Warsaw: Foundation for the Development of Polish Agriculture; 2010, p95-120.

[72] BirdLife International. Agri-environment schemes and biodiversity: lessons learnt and examples from across Europe. http://www.birdlife.org/eu/pdfs/Agrienvironment_schemes_lesson_learnt.pdf (accessed: 8 February 2012).

[73] Paloniemi R, Tikka PM. Ecological and social aspects of biodiversity conservation on private lands. Environmental Science and Policy (ISSN: 1388-0292) 2008;11 336-346.

[74] Knight AT, Cowling RM, Difford M, Campbell BM. Mapping human and social dimensions of conservation opportunity for the scheduling of conservation action on private land. Conservation Biology (ISSN: 0888-8892) 2010;24 1348-1358.

[75] Herbut E, Walczak J. Ekstensywne rolnictwo w Polsce: conference proceedings, 18-27 February 2008, XV science conference. Postęp naukowo- techniczny i organizacyjny w rolnictwie, Zakopane, Poland. 2008.

[76] Larsson TB. Biodiversity Evaluation Tools for European Forests. Ecological Bulletins (ISBN-10: 8716164342) 2001;50.

[77] Niemelä J. Identifying managing and monitoring conflicts between forest biodiversity conservation and other human interests in Europe. Forest Policy and Economics (ISSN: 1389-9341) 2005;7(6) 877-890.

[78] Fawcett SB, Paine-Andrews A, Francisco VT, Schultz JA, Richter KP, Lewis RK, Williams EL, Harris KJ, Berkley JY, Fisher JL, Lopez CM. Using empowerment theory in collaborative partnership for community health and development. American Journal of Community Psychology (ISSN: 0091-0562) 1995;23(5) 677-697.

[79] George W.E., Mair H., Reid D.G. Rural tourism development: Localism and cultural change. Bristol, Buffalo, Toronto: Channel View Publications; 2009.

[80] Wilkinson KP. The community as a social field. Social Forces (ISSN: 0037-7732) 1970;48(3) 9-17.

[81] Pratchett L, Durose C, Lowndes V, Smith G, Stoker G, Wales C. Empowering communities to influence local decision-making: A systematic review of the evidence. Department for Communities and Local Government 2009. http://www.communities.gov.uk/documents/localgovernment/pdf/1241955 (accessed 01 February 2012).
[82] Zimmerman MA, Rappaport J. Citizen participation, perceived control, and psychological empowerment. American Journal of Community Psychology 1988;16(5) 725-750.
[83] Li WJ. Community decision-making participation in development. Annals of Tourism Research (ISSN: 0160-7383) 2006;33(1) 132-143.
[84] Cole S. Information and Empowerment: The Keys to Achieving Sustainable Tourism. Journal of Sustainable Tourism (ISSN: 0966-9582) 2006;14(6) 629-644.
[85] Tosun C, Timothy D. Arguments for community participation in tourism development. Journal of Tourism Studied (ISSN: 0974-6250) 2003;14(2) 2-11.
[86] Bora A, Hausendorf H. Participatory science governance revisited: normative expectations versus empirical evidence. Science and Public Policy (ISSN: 0302-3427) 2006;33(7) 478-488.
[87] Knopp TB, Caldbeck ES. The role of participatory democracy in forest management. Journal of Forestry (ISSN: 0022-1201) 1990;88(5) 13-19.
[88] Armitage DR. Community-cased Narwhal management in Nunavut, Canada: Change, uncertainty, and adaptation. Society & Natural Resources (ISSN: 0894-1920) 2005;18(8) 715–731.
[89] Austin RL, Eder JF. Environmentalism, development, and participation on Palawan Island, Philippines. Society & Natural Resources (ISSN: 0894-1920) 2007;20(4) 363–371.
[90] Antoniewicz P. Partnerstwo czlowieka i przyrody, Dolnośląska Fundacja Ekorozwoju. http://www.iee.org.pl/rozwoj/docs/PARTNERSTWO_CZLOWIEKA_I_PRZ.pdf
[91] McCauley D. Sustainable development and the governance challenge: the French experience with Natura 2000. European Environment (ISSN: 0961-0405) 2008;18(3) 152-167.
[92] Agrawal A, Gibson G. Enchantment and disenchantment: The role of community in natural resource conservation. World Development (ISSN: 0305-750X) 1999;27 629-649.
[93] Walker PA, Patrick T, Hurley PT. Collaboration derailed: The Politics of "community-based" resource management in Nevada County. Society & Natural Resources (ISSN: 0894-1920) 2004;17(8) 735–751.
[94] Reid DG, Mair H, George W. Community tourism planning: A self-assessment instrument. Annals of Tourism Research 2004;31(3) 623-639.
[95] Lachapelle PR, McCool SF, Patterson ME. Barriers to effective natural resource planning in a "messy" world. Society & Natural Resources (ISSN: 0894-1920) 2003;16(6) 473–490.
[96] Smith PD, McDonough MH. Beyond public participation: Fairness in natural resources decision making. Society and Natural Resources (ISSN: 0894-1920) 2001;14 239 -241.
[97] Stec S., Casey-Lefkovitz S., Jendrośka J. (editors.) The Aarhus Convention - An Implementation Guide. New York and Geneva (ISBN: 9211167450 9789211167450) 2000: Channel View Publications.
[98] Dziennik Ustaw - year 2003, nr 78, item 706 from day 2003-05-09.

Chapter 10

Economic Valuation as a Framework Incentive to Enforce Conservation

Isabel Mendes

Additional information is available at the end of the chapter

http://dx.doi.org/10.5772/50063

1. Introduction

A Protected Area (PA) is an "area of land and/or sea especially dedicated to the protection and maintenance of biological diversity, and of natural and associated cultural resources, and managed through legal or other effective means" [1]. In this paper biological diversity means "...the dynamic network of biological, chemical, and physical interactions that sustain a community and allow it to respond to changes in environmental conditions" (in [2], p. 41). Consequently PA means biological diversity conservation, or ecosystem conservation as a whole, with their geographical, biological and different types of resilience to different types of human activities resulting pressures. By adopting this definition we will be following the Subsidiary Body on Scientific, Technical and Technological Advice of the Conference of the Parties to the Convention on Biological Diversity that indicates the importance of looking at biodiversity conservation under an ecosystem approach rather than focusing on the individual components within the ecosystems.

PA's are the groundwork of conservation policies [3-7], and therefore, traditional biological conservation management practices have been based on them [7-8] and on the Safe Minimum Standard Principle (SMS), in line with the ecological perspective of sustainability as defined by the Daily Rule [9]. The SMS principle states that a sufficient area of ecosystem must be conserved to ensure the continued provision of ecosystem services unless the social costs of conservation are unacceptably high. The Daily Rule states "never reduce the stock of natural capital below a level that generates a sustained yield unless good substitutes are currently available to the services generated" (In [9], p. 112, citing Daly H. (1996) Beyond Growth: The Economics of Sustainable Development. Boston: Beacon Press. 81-82).

Nevertheless, PA's based conservation policy by itself seems not to be sufficient to achieving the required conservation levels for outstanding ecosystems, with resident populations, located near urban centres and with good means of access and especially within a low or middle income country context [10]. Over the last century there has been a growing realisation that biodiversity is being lost at an alarming rate, in spite of the national efforts to expand protected areas to incorporate a wide variety of ecosystems put into danger by the human population expansion, increasing damaging pressure placed by human activities like deforestation, pollution, poaching, or intensification of agriculture. One reason for such insufficiency is that biodiversity in general and PAs in particular, exist neither in isolation nor are independent of human activity. Over 50% of the 20,000 official PAs established over the last 200 years are on land historically occupied by indigenous peoples [8]. Furthermore, some stress that official conservation initiatives have tended to neglect some of the subsequently protected environments as the result of their long-term interaction with humans. This is surely the case of Europe, and Portugal in particular. In these regions, landscapes encompass large areas of semi-natural vegetation interspersed with grazing areas, hedgerows, farmland, and small villages and towns. Or, as in the case of wetlands, the coastline is frequently host to important urban communities depending on fishing activities and/or other marine activities and/or tourism.

Despite the richness and historical record of such human environmental interaction, earlier conservation practices have implemented an ideology primarily based on nature and some kind of "conservation ethic" potentially devaluing those values and systems that sustained human ecological practices in their respective contexts. Currently, while most of the governmental conservation agencies broadly recognise and accept the notion of "conservation-with-use" [8] by explicitly integrating it into national conservation legislation most state-declared PA local communities have not been active participants in designating or managing their surroundings. This state-imposed development and conservation, leaving local community on the peripheries of power, has created local antagonism and suspicion [8]. PA inhabitants, be they local farmers, herders or fisherman, all complain about being marginalised. When the state sets aside PAs, residents claim it does so at the cost of their own land, resources, and livelihoods, mostly without proper compensation for loss of property or forgone revenues [11]. The act of conservation is thus interpreted as a misfortune rather than an opportunity for sustainable development, accordingly to the Triple Bottom-Line methodology. This thereby leads to increasing evasion of conservation regulations while the government delays in answering the worsening conservation problems resulting from those individual actions [12,13].

Although current economic investigation is proving that enforced resource conservation measures like PA's are efficient in the sense that they do not cause Pareto inefficiency and that Pareto optimal allocations cannot only be reached through competitive markets [14], we would additionally state conservation practices have to change sharply, and must stretch beyond the SMS principle through both engaging locals and other users with the conservation process and creating a broad consensus as to the existence and objectives of conservation initiatives.

Within this framework, Co-Managed Protected Areas (CMPA) is emerging. CMPA are official state-established PAs managed with the effective engagement of other social actors, including indigenous and local communities [8]. They are universally accepted means of community enforcement [8] with three objectives: the conservation of local natural and cultural heritage, the participation of civil society in the management process and the equitable distribution of benefits and costs. Community empowerment has to be reinforced with the implementation of incentive measures to enforce actor compliance, reinforce capacity building and provide intelligible information about the value of nature conservation (for example, see [2, 12] for a comprehensive description about incentive measures for biodiversity; or [15] as an example of how to use a travel cost approach to estimate the value of an entrance fee for a National Park). One of the means of achieving the aforementioned community empowerment is to make people aware that a PA is a sort of capital, - the natural capital -, and is as valuable as any other sort of capital goods like houses, land, art, factories, or infrastructures being able in generating flows of monetary benefits. Following [16], natural capital is generally defined as the stock of environmentally provided assets (e.g. soil, atmosphere, forests, water, wetlands, minerals) generating flows of goods and services, that are appropriate by economic sector and society at free cost. Locals must be aware that setting land aside by government for conservation purposes, instead of being a curse, may be a way of earning money and meliorating their way of living. The key idea is to ensure PA inhabitants as the owners of the natural capital, gain a strong economic interest in protecting that capital and therefore to guarantee the sustainability of the environment they live in, either for their own profit or to maintain resale value [9]. This is by no means an easy task for policy-makers and is packed with social, scientific, and practical difficulties and ambiguities [17-18]. Furthermore, biodiversity policy has its own respective complexities, differing to classical pollution problems, including heterogeneity, irreversibility, accumulation of impacts, information gaps, mix of values and pressures [2]. Since economic decisions are market-based, it may be necessary to apply economic instruments to conservation policy, like economic incentives, prices, or information concerning the monetary values of the benefits generated by the natural capital, and the monetary value of the natural capital stock itself. This is profit-based conservation and we defend its application in conjunction with the SMS principle to improve conservation policy rather than as any simplistic alternative. As several economists and ecologists have been warning, too little nature will be conserved by market force alone [13, 19].

In this chapter, we discuss the advantages and disadvantages of economic valuation of ecosystems as an incentive measure to enforce local community co-operation in conservation decisions and management. This issue is not new to economic and ecological literature. However, the literature mostly serves to demonstrate that economic valuation is anything but exhausted as an issue for discussion. Misleading pro and contra arguments and definitions are often used, both by economists and ecologists, making any understanding of monetary evaluation of biodiversity and its respective methodology more complicated than is justifiable. By recognising such difficulties, this chapter tries to assess the main points of discussion around the economic valuation of biodiversity, including the advantages and disadvantages of applying it as an incentive tool to enforce conservation attitudes. Analysis is only provided from the economist's point of view and does not seek to be exhaustive. The chapter is organised as follows. We proceed by clarifying what "value" means to economics. In section 3, we describe the most important steps required for ascertaining the most accurate possible monetary value. In section 4, we describe why monetary valuation is an important and reliable tool in policy and conservation management despite these difficulties and controversies. Finally, conclusions are drawn.

2. The value of ecosystems: What does this mean to economics?

In common usage, "value" means importance or desirability. To an economist, an ecosystem is a non-marketed good, and therefore its value is related to the contribution it makes to human wellbeing (see for instance [20-21] to read more about the state of non-market valuation of environmental resources). Human wellbeing depends on the basic requirements for a good quality of life including freedom of choice, health, good social relations and security and may be broadly understood as happiness. Wellbeing, as experienced and perceived by humans, is situation-dependent as it reflects the local geography, culture and environmental circumstances. We are dealing with a very clear anthropocentric, utilitarian viewpoint according to which ecosystems are valuable insofar as they serve humans or to the extent they confer any sort of satisfaction on humans [21, 22].

The utilitarian approach allows value to arise in a number of ways depending on how individuals use ecosystems [21, 23]. The "prior" or "primary value" consists of the system characteristics upon which all ecological functions depend (resilience capacity, individual resource stability, biodiversity retention). The value arises in the sense that the ecosystem produces other functions with value – "secondary functions", and, as such, in principle, has economic value. These secondary functions and associated values depend on the maintenance, health, existence and the operationality of the ecosystem as a whole. Hence, economists have generally settled for taxonomy of total ecosystem value interpreted as a Total Economic Value [23] (TEV) that distinguishes between Direct Use Values and Passive (Non-use) Values. Passive Use is now used interchangeably with Non-use or Existence Value. Other terms that have been used include preservation value, stewardship value, bequest value, inherent value, intrinsic value, vicarious consumption and intangibles [24]. Figure 1 provides a diagram detailing the relationship between the TEV's taxonomy of use and non-use values and the ecosystem's service concept.

Use Values include Direct Use Values and Indirect Use Values (see [2, 25-26] for a more detailed definition of the different type of uses). Direct Use Values derive from the actual use (consumptive and non-consumptive) of natural resources for: commercial or self-consumption purposes (e.g. harvesting timber, fishing, collecting herbs and minerals); tourism and recreation; education and research; aesthetic, spiritual, and cultural ends. These are the so called primary functions of the ecosystems. One of the more recognised and important sources of ecosystem's use value, is that associated to knowledge. According to a report for the U.S. National Academy of Sciences, the basic source of over $60 billion in current market value was obtained from biodiversity (plants and insects). Furthermore,

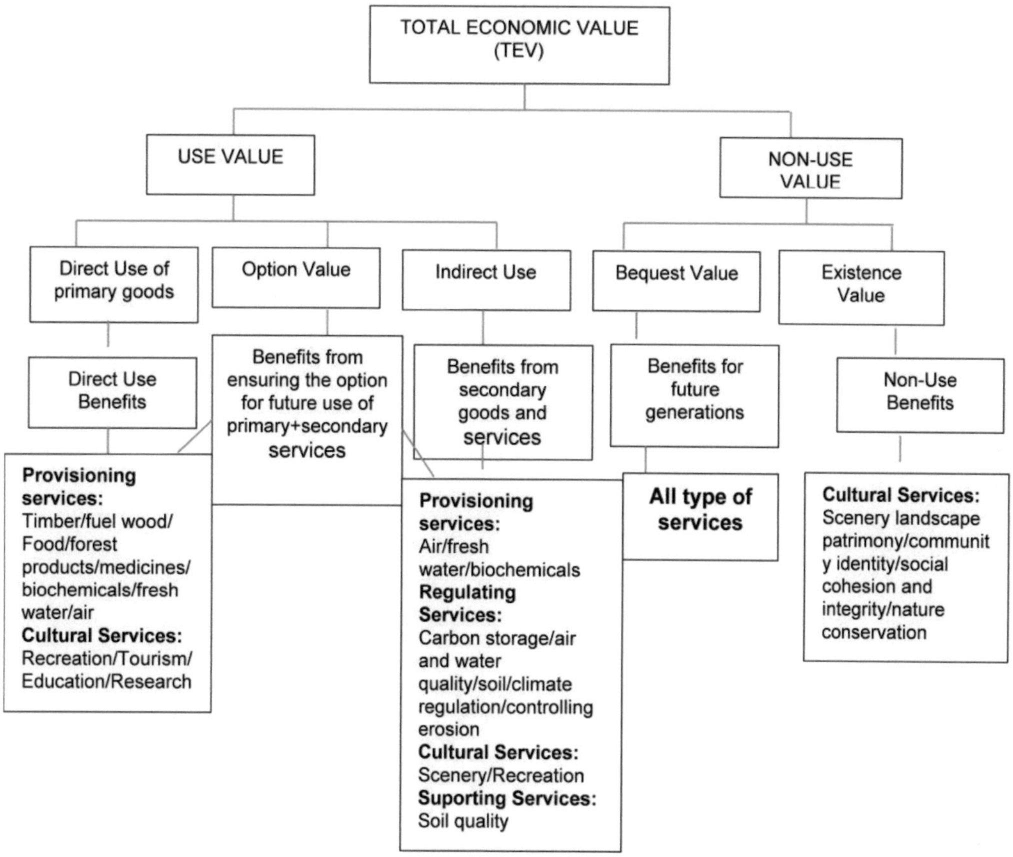

Figure 1. TEV Taxonomy and Ecosystem Services (adapted from [25])

several of the most productive and robust grain species were genetically derived from wild specimens. Currently, most drug and biotechnological companies are well aware of the strategic, commercial and scientific value of ecosystems as vital depositories of the strategic, commercial and scientific value of ecosystems (see for instance [27] for more details and [18] for their interesting parabolic perspective of ecosystem and biodiversity in terms of Noah's Ark). Indirect Uses are related with the use society makes from ecosystem functions (or secondary functions) indirectly, like watershed values (e.g. erosion control, local flood reduction or regulation of stream-flows) or ecological processes (e.g. fixing and cycling nutrients, soil formation, cleaning air and water). Indirect Use values may further include "Vicarious Use value" addressing the possibility that an individual may gain satisfaction indirectly, from pictures, books, or broadcasts of natural ecosystems even when not able to visit such places. Option Values are related with individual willingness to pay a premium to ensure future ecosystem availability and usage. Some authors also refer to a Quasi-Option Value which reflects the individual willingness to pay a premium to ensure the maintenance of the ecosystems and respective services to get more accurate scientific information in the future. This type of benefit relates with individual perceptions concerning the irreplaceability degree of some ecosystems and with the lack of information about its functioning. Given this type of uncertainty, some individuals prefer to conserve ecosystems instead of destroying them until the moment society will have more accurate information about their real ecological value. Passive (or Non-use) Values include Existence Value: reflects the moral or altruistic satisfaction felt by an individual from knowing that the ecosystem survives, unrelated to current or future use. Finally, Bequest Value considers individual willingness to pay a premium to ensure that their heirs will be able to use the ecosystem in the future.

TEV and its components has been the subject of huge debate among environmental economists, ecologists, psychologists, and others, about the viability, the usefulness, or the ethics of monetising it, especially passive uses (for a more comprehensive understanding of this debate see for instance [23]. Nevertheless there is actually a growing trend towards using the TEV measure on the grounds that theoretically there is no need to adopt a dichotomy that involves the adoption of arbitrary assumptions. See [28- 29] for a more comprehensive understanding of the total and non-use values discussion. Advances in ecological economic models and theory also seem to stress the value of the overall system as opposed to individual system components only. Currently, ecosystems or environmental resources as a whole are increasingly recognised as assets providing sets of services that are no longer readily available. Increasing demand to measure their value and incorporate them into legal, political, and economic decisions is a clear sign of what we would expect as their scarcity grows [20-21]. This point to the value of the system itself when exhibiting resilience capacity defined as the ability of the ecosystem to maintain its properties of self-organisation and stability while enduring stress and shock [23].

The economic value of an ecosystem to some individual, thus relates to TEV that individual puts on the ecosystem. Individual's TEV is not an absolute value because economics provides valuations only in comparative terms. When economists say they are valuing an ecosystem, they are really defining a trade-off between two situations involving a change: e.g. maintenance or non-maintenance of the ecosystem. The economic value of the ecosystem is the amount an individual would pay or be paid to be as well off with the ecosystem or without it [30-31]. Economic value is therefore an answer, mostly expressed in monetary terms (although not necessarily), to a carefully defined question in which two alternatives are being compared. The answer (the value) is very dependent on the factors incorporated in that choice: the object and the circumstances of choice [32]. Economics defines objects of choice as any tangible or non-tangible object, process or activity that can be described as allowing choice and are defined by a set of characteristics and attributes that are perceived by individuals but not necessarily by all individuals. In our case, the object of choice is an ecosystem whose specificity is defined by a set of environmental and ecological attributes to a greater or lesser extent perceived by individual users and passive users. The circumstances of choice describe the context in which that choice is made (to accept or not the political option to conserve the ecosystem). It is clearly fundamental to describe to the individual the consequences of his/her choice, specifically in terms of: i) what is foregone by the choice and what is gained; ii) specify the rights of assignment; iii) define the mechanism of choice, that is the manner through which the individual will exercise choice: by voting, through private market transactions or other unspecified behaviours.

The object and circumstances surrounding such choice define its context. In the case of ecosystems, value depends on the ecosystem location and the level of human presence, the actual or threatened level of degradation as well as the degree to which natural services provided can be substituted by other substitute ecosystems. This substitutability is a highly important concept within economic valuation as objects with significant numbers of close substitutes are not rated as valuable as those with few or even no substitutes. In the case of ecosystems, the degree of substitutability is relative depending on factors including the scale and level of aggregation and the time-scale involved. For specifying rights of assignment, there are two possible choice situations. Either the individual gives something up to receive the object of choice that will affect his/her utility or well-being or the individual receives something to give up the object of choice that could affect his/her utility or well-being. The former situation corresponds to Willingness to Pay (WTP) and the latter to Willingness to Accept (WTA) and these are the fundamental monetary measures of value in economics.

These welfare measures applied to non-market transacted objects of choice were first proposed by Mäler [33-34] as an extension of the standard theory of welfare measurement related to market price changes formulated by Hicks [35]. Mäler stated that it was possible to build four measures of individual welfare change associated to choices involving non-market goods. If the object of choice (for instance, a conservation policy) generates an improvement in individual well-being (that is a rising utility), two situations become possible. Either the individual is WTP an amount to secure that change, termed Compensated Willingness to Pay (WTP^C) or he/she is willing to accept a minimum of compensation to forgo it, the Equivalent Willingness to Accept measure (WTA^E). If the object of choice generates a deterioration (for instance, some ecosystem destruction, related with a particular human activity) in well-being (that is, a decreasing utility), again two situations are possible. Either the individual is WTP to avoid this situation, termed the Equivalent Willingness to Pay measure (WTP^E) or he/she is WTA compensation to tolerate the damages suffered, the Compensated Willingness to Accept measure (WTA^C). When economists talk about the value of an ecosystem they are referring to an individual TEV measured by one of these four welfare measures: WTP^C/WTA^E if the individual faces an improvement of wellbeing; or WTP^E/WTA^C where the individual faces deterioration in well-being.

Mäler used the following basic model of individual utility to define welfare measures. Let $U = U(x,q)$ be the utility function of an individual with preferences for various conventional market commodities and where consumption is denoted by the vector $\mathbf{x}$, and for non-market environmental amenities denoted $\mathbf{q}$. $\mathbf{q}$ may be a scalar where related to a single amenity or is a vector where related to several amenities as is the case of $\mathbf{q}$ representing the ecosystem one wishes to value. The individual takes $\mathbf{q}$ as given which means $\mathbf{q}$ is a public good. It is also assumed that preferences represented by the utility function are continuous, non-decreasing and strictly quasi-concave in $\mathbf{x}$. The specific form of the utility function will affect the shape of the indifference curves. The shape of the indifference curves indicates the preferences the individual has for **x** and **q**. In this case, to say the utility function is quasi-concave is merely for the sake of analytical convenience. To say this is realistic or not is considered by utilitarian theory as being a merely empirical question [36]. The individual faces a budget constraint based on their disposable income m, and the prices of market commodities, **p**. The individual maximisation utility problem of decision is then formalised as:

$$\max_{X^*} U(x,q) \quad \text{subject to} \sum p_i x_i = m \tag{1}$$

meaning, each individual is trying to decide what is the affordable quantity of good X (x*) that must be purchased in order to get the maximum level of utility (or satisfaction), given his/her per capita income, the prices of the good, and the level of amenity q, all remaining constant. The solution of problem (1) yields a set of ordinary or marshalian demand functions for **x** denoted $x_i = g_i(p,q,m)$, for i = 1, ..., N individual and an indirect utility function denoted $U(x,q) = \varphi(p,q,m) = U\left[g_i(p,q,m);q\right]$. Each individual marshalian demand function expresses the relationship between the changes on the individual purchases of good X as its price rises, holding individual income and the amenity constant; and each indirect utility function expresses the changes on each individual utility (or satisfaction), for different purchases of the good X. The dual problem of each individual i is an expenditure minimisation model defined by:

$$\min_{x_c} \sum_{i} p_i x_i \quad \text{subject to} \quad U(x,q) = U \tag{2}$$

The dual problem states that individual i wants to make the lowest possible expenditure (denoted $p_i x_i$) with X, to maintain his/her utility or satisfaction at a particular level denoted U, for all level of prices. The solution of the dual problem (2) yields a set of compensated or Hicksian demand functions for **x** denoted $x_{ic} = h_i(p,q,U)$, and an expenditure function (3):

$$m = e(p,q,U) = \sum_{i} p_i h_i(p,q,U). \tag{3}$$

Each compensated demand function shows how the quantity demanded of good X changes as the price rises, holding utility and the amenity q unchangeable. The expenditure function shows the relationship between the minimal expenditures with the good X necessary to achieve the specific level of utility.

Let us now suppose **q** is going to change from the current state $\mathbf{q}^0$ to another different state $\mathbf{q}^1$ so that the individual i will have to choose between the state $\mathbf{q}^0$ or the state $\mathbf{q}^1$, while holding constant the individual per capita income m and the prices p. If he or she chooses $\mathbf{q}^0$, the level of utility is given by $U^0 = U^0(p,q^0,m)$; if he or she chooses $\mathbf{q}^1$ the utility is given by $U^1 = U^1(p,q\ \ ,m)$. The welfare change associated with the change of the utility level from U^0 to U^1 can be measured using Mäler's Compensation Variation (CV) or Equivalent

Variation (EV) measures, defined respectively by $CV = e(p,q^0,U^0) - e(p,q^1,U^0)$ and $EV = e(p,q^0,U^1) - e(p,q^1,U^1)$, as they are illustrated in Figure 2.

Hicksian measures \ Utility change	Ecosystem improvements $q^0 < q^1 \Rightarrow U^0 < U^1$	Ecosystem degradation or destruction $q^0 > q^1 \Rightarrow U^0 > U^1$
CV	WTP^C	WTA^C
EV	WTA^E	WTP^E

Figure 2. The relationship between the Hicksian CV and EV utility measures and the WTP and WTA Malher's utility measures

If $U^1 > U^0$, that is if there is an improvement of the individual's utility or level of satisfaction associated with an improvement of the ecosystem, utility CV measure are WTP^C to secure that change, or EV measure WTA^E to forgo it. If $U^1 < U^0$, that is if there is a decrease of the individual's utility or of the level of satisfaction associated with the depreciation or destruction of the ecosystem, utility CV measure are WTA^C to tolerate the damage, and EV measure WTP^E to avoid that change. Given the duality between the indirect utility function and the expenditure function, the fundamental monetary measures are, therefore, given by equations (4) and (5), respectively:

$$WTP^C or\ WTA^C = CV = e(p,q^1,U^0) - e(p,q^0,U^0) = \int_{q^0}^{q^1} \frac{\partial e(p,q,U^0)}{\partial q} \quad (4)$$

$$WTP^E or\ WTA^E = EV = e(p,q^1,U^1) - e(p,q^0,U^1) = \int_{q^0}^{q^1} \frac{\partial e(p,q,U^1)}{\partial q} \quad (5)$$

In short, when environmental change is positive and individual's utility rises, he/she will be able to pay a maximum amount of money given by WTP^C to secure the environmental improvement, or he/she will accept a minimum amount of money given by WTP^E to forego the environmental improvement. When environmental change is negative and individual's utility declines, WTA^C is the minimum amount of money that must be given to the individual to compensate him from the environmental damage effects and WTP^E is the maximum amount of money the consumer has to pay to stop the implementation of the economic decision that will be the cause of the environmental damage.

WTP (equivalent or compensated) and WTA (equivalent or compensated) may or may not produce differing monetary information for the same object under valuation. The difference between the two measures is explained by the income elasticity (that is the sensitivity of the quantity demanded at a given price to income changes) and the budget share of the ecosystem. The smaller the income elasticity or the smaller the budget share, the closer are the measures to each other. WTP is equal to WTA where there is no income effect and individual is neutral to losses and gains, and where there are perfect substitutes of the object under valuation. Generally, WTA is greater than WTP which stems from the different welfare measure definitions and contexts of choice: as proved by numerous empirical estimations of WTP/WTA, WTA is always greater than the WTP because the latter is limited by the individual's budget restriction and by the existence or non-existence of substitutes, while the former is not. See [37-38] for a more comprehensive discussion about the WTP and WTA differences and the consequences they have on the valuation of environmental services.

For purposes of economic valuation, the ecosystem and its components (q) are considered to be public or quasi-public reproducible natural assets, like structures or equipment. By the fact of being natural assets, they have direct and indirect use values, and passive uses as well, including option, quasi-option and existence values. As people experience satisfaction (utility) with the existence and services of ecosystems, the value of that reproducible natural asset is necessarily linked to its generated value flows of services and passive uses (it is assumed that the satisfaction or utility $U(q^1)$ generated by the new state of the ecosystem is a flow that will occurs during a particular time period of time). Hence, according to the asset analogy, TEV is equal to the discounted sum of values (or benefits, or utilities) of those services and passive use benefit flows, so that:

$$TEV = \sum_{t=0}^{T} \frac{TEV^t}{(1+\rho)^t} \tag{6}$$

where T is the relevant period of time during which the ecosystem generates the flow of amenities; TEV^t is the mean WTP (or WTA) each individual wants to pay for the amenity's improvement (or be compensated for the destruction of the amenities) estimated for the relevant population (that is, for the individuals that are going to suffer the effects of the amenity's change) at a t point period of time period T; and ρ is the discount rate used to discount the cash flows provided by the ecosystem during the T time period considered.

3. Estimating ecosystems TEV

WTP and WTA are though the fundamental monetary measures of value in economics for market and non-market commodities. When economists set about estimating the individual's WTP/WTA (or the value) for the market goods and services, they use actual, observed, market-based information, because preferences for private goods are revealed directly when individuals purchase them on the market. However, because ecosystems and most of the services they produce, like the ones associated with indirect use, option or existence values, are not market tradable, we thus have to elicit individual preferences directly in terms of WTP/WTA by use of questionnaires, such as Contingent Valuation (CV). CV provides the means to estimate natural resource value or loss and is the only current method that produces estimations of the ecosystem's TEV, including all sort of the marketed and non-marketed ecosystem services. For a detailed description of the Contingent Valuation method, see for instance [39] or, more recently, [40]. If you want to apply the method in order to estimate the value of some ecosystem or the value individuals put on some ecosystem's changes, you may consult [39-41]. For a more detailed description of other economic valuation methods, see for instance [38, 42-43]. So, to estimate (6), one has to go through the following steps: i) to estimate WTP/WTA and TEV via actual individual responses gathered by the CV method; ii) to estimate TEV, using aggregate values of stated WTP/WTA discounted by a certain rate, in accordance with the natural asset analogy.

3.1. The use of the CV method and estimate reliability

The use of CV methods to estimate TEV for environmental services has been one of the most fiercely debated issues within environmental economic valuation literature until the 90's of the last century, more specifically the validity and reliability of CV estimates issue and the inclusion or non-inclusion of passive users as a TEV component [44,16]. Detractors argue that respondents provide answers inconsistent with the basic assumptions of utilitarian rational choice; they question the seriousness of CV answers because the results of surveys are not binding. And a more extreme position even holds that the economic concept of value itself has no link to reality. This has led to the supposition that responses might bias CV value away from theoretical welfare measures. Defenders acknowledge that early applications suffered from many of the problems critics have noted [39]. However, recognition is required of how more recent and more comprehensive studies have dealt and continue to deal with those objections.

The key to a successful evaluation method is that it must be assessed in terms of how closely it represents an accurate measurement of the real value. The closer the real values are to the estimated, the more accurate the valuation method is. If WTP/WTA an ecosystem were observable there would be no problem. But given they are not, it is then necessary to use other complex criteria and "rules of evidence" to assess accuracy. In measurement, accuracy means the reliability and validity of data analysis used for the valuation framework. See [39-40] for a more comprehensive description of these methodological CV problems; and their potential effects upon estimates. See also [16] for a comprehensive survey of literature on such issues.

From the economics perspective, reliability is related with the accuracy of aggregate WTP over appropriately defined aggregates of individuals. Economists tolerate certain amounts of unreliability in the estimated WTP, if random errors in measurement remain within tolerable boundaries. The bias between the theoretical, and the CV estimated WTP/WTA, grows where the latter tends to systematically diverge from the former. The concept validity relates to the CV application process that involves numerous issues that must be resolved mainly based on individual judgement of the CV implementing entity. Because WTP is not observed, inferences as to validity are based on indirect evidence related both with content validity of CV study design and execution, and construct validity dealing with the degree to which the estimated money measure relates to other theoretical measures. To assess the content validity involves examining study procedure content. This involves four steps. Firstly, the researcher defines the scenario that would lead a theoretical consumer to reveal his or her WTP/WTA. More precisely, this is the CV phase where the elements of choice or details of the transaction are presented to the respondent. The transaction must be adequately defined in order to be clearly understood by the participant. The second step is vital to controlling the extent to which participants really understand the proposed transaction as communicated through the scenario defined in the first step. This obliges the introduction of qualitative research procedures to support CV survey design. The third and fourth steps refer to the appropriate statistical and econometric techniques that must be applied to elicit unbiased, higher content validity estimates of WTP/WTA. Contributions towards improving CV studies, from both the theoretical and empirical perspective, have been drawn from the differing fields of academic social science research – economics, psychology, law and politics (see [32] for a more comprehensive study of this issue) - but the most important was that of NOAA Panel Report [45]. Recognising the impossibility of externally validating estimates produced by CV studies, the NOAA Panel recommended researchers adopt an *ex ante* analysis of the results in place of an *ex post* analysis, by focusing discussion on how to improve the theoretical and empirical quality of studies thereby improving the accuracy of CV valuation by strengthening result reliability. The Panel guidelines for the study design phase are set out defined for three aspects of the transaction: the good, the payment, and the valuation context. In the case of ecosystem valuation surveys, respondents need to know about the attributes of the ecosystem [46-48], the level of provision of those environmental attributes "with and without intervention" and if there are undamaged substitute commodities. As for payment, the Panel recommended the use of the WTP valuation format and the definition of the "payment vehicle" which may include taxes, property taxes, sales taxes, and entrance fees, changes in the market prices of goods and services or donations to special funds. As for the transaction context, it is important to explain the extent of the "market" by informing respondents of how and when the environmental change will occur as well as the decision rules in use for such provision (e.g. majority vote, individual payment). Researchers must allow respondents the opportunity not to vote. The Panel recommends a conservative survey design, a referendum style choice, and voting choices followed by open-ended questions asking about reasons for voting one way or another. The Panel's recommendations on survey design are for the most part almost standard practice except: i) the use of a referendum format in substitution of the open ended question eliciting the maximum WTP; ii) and the opportunity to the respondent to choose not to vote. These Panel guidelines are standard practice for any high quality survey. It is recommended to use probability sampling, in-person interviewing, to minimise non-responses, to make careful pre-testing and to examine interviewer effects.

By recognising the impossibility of independent verification of the CV results, the Panel suggested that besides the survey design and administration guidelines, an alternative test of CV reliability must be drawn from the economic theory of rational choice to monitor the rationality of individual WTP responses. This test is called the scope (embedding) test and requires that the stated survey WTP should be related to the size of the object of choice. If the WTP response is inadequate to the scope of the object of choice then the findings of the CV survey are unreliable. The scope test is based on the weakest form of rationality among individual choices. It is reasonable to suppose that more of a good is always better to the individual if not satiated and that he/she is willing to pay more for more of that good. Also, it is reasonable to assume that WTP will decline although not abruptly for additional amounts of the good [45]. The NOAA Panel concluded that the information provided by CV surveys, where in full compliance with the recommendations, can be considered "as reliable by the standards that seem to be implicit in similar contexts, like market analysis for new and innovative products, and the assessment of other damages normally allowed in court proceedings" [45].

More recently there has been a trend to include expertise from other disciplines such as marketing research, survey research, cognitive and social psychology (see [46] for a comprehensive introduction to the psychological perspective on economic valuation to improve the CV methodology) both from the theoretical and empirical point of views. To read more about psychologist's criticism of the utilitarian approach to ecosystem valuation, see for instance [49] , where the problem related with the existence of lexicographic preferences is discussed, and [50] , for a review of the some empirical evidence of such preferences for environmental goods. The importance of all these contributions to survey research is almost intuitive because CV is broadly survey valuation method based.

3.2. Estimating TEV using individual stated WTP

As mentioned earlier in the chapter, the capital asset pricing approach views the ecosystem value as an asset value (natural and reproducible) at a particular point in time as the discounted value of all future services the ecosystem will provide as stated by equation (6). A common economic approach is to assume a rate of discount and further assume that the flows of services provided by the ecosystem during the relevant period of time T will be constant, and that the value of flows will increase in line with the expected rate of inflation. Under these assumptions, one is left with the task of valuing the services at some point in time t discounted at some discount rate, ρ. The problem here is what rate of discount is to be chosen and what flows are to be considered.

The answers to either question are not straightforward to economists because the valuation of ecosystems or the decisions involving nature related sustainability decisions are characterised by several dimensions in which issues are more demanding for economists than those raised by environmental economics. One is the time dimension. The long-term period T applied to ecosystems, not inferior to 50 years and sometimes as long as one, two or even several centuries, is much longer than normally considered in economic analysis, which poses a particular challenge for the economist's traditional practice of discounting.

A second dimension is related with uncertainty. Over such a long time scale, it is logical to expect that individual preferences will change due to technical and social changes that also affect the flow of individual WTP across time. On the other hand, society's ecosystem services demand is already so great that trade-offs among competing services have become a rule. This increasing demand is compounded by the increasingly serious degradation of ecosystem capacities to provide services, seriously diminishing the prospects for sustainable human development [51]. It is though important to incorporate the ecosystem TEV into this pressing trend for scarcity. As one cannot ask people to be futurists as to their future WTP patterns, one way of achieving this is to use appropriate discount rates and discount methods that allow for incorporating growing future ecosystem scarcity. The default criterion that has been used for ranking environmental conservation projects is provided for by the discounted utilitarian approach introduced by Bentham in the nineteenth century, and has being dominating more due a lack of convincing alternatives rather than any intrinsic accuracy. According to it, one must choose the greatest present discounted value of net benefits. The project must be rejected where it provides a negative net benefit. A positive utility discount rate ρ reflects what economists refer to as a positive time preference [52], a widespread desire to consume today rather than save for the future. This is a way by which the market penalises investments with long-term payoffs: any investment with high up-front costs and a long stream of future benefits will dramatically undercount future benefits. This is precisely what happens when one has to value ecosystems by discounting the flow of the services they provide over time: at any positive discount rate, the present value of any ecosystem is almost irrelevant and it thus becomes irrational (from the economic point of view) to be concerned about extinction or conservation. And yet societies obviously are worried about such issues and actively continue to consider how to devote substantial and scarce financial resources to them.

How important then is our concern as economists about the time dimension to ecosystem valuation frameworks? The opinion of some authors like Ramsey [53] or Harrod [54] (the original proponents of modern dynamic economics), and that of Heal [55] on environmental subjects, is that discounting is ethically indefensible (see chap. V in [18], for a comprehensive theoretical approach to this issue). There is however empirical evidence suggesting the legitimacy of discounting future utilities even where it is not certain that discounting future utilities in the evaluation development programmes is the same as discounting benefits of values [55]). Where empirically proven that people consider some positive discount rate we have to conclude that the traditional discounted utilitarianism approach to value benefits is not particularly suitable to clearly (as far as possible) capturing individual future concerns over the future environment. This is because decision-makers and cost-benefit applying researchers generally use market rates of discount much higher than the appropriate efficient sustainable rate thus depreciating future flows. As a consequence of this controversy, researchers have, in some cases, begun to apply lower discount rates to long-term, intergenerational projects [56]. Others use a declining rate in the future. Unfortunately, both methods result in time-inconsistency problems as long-term projects in the present become near-term projects in the future (see [18] for a more comprehensive discussion of this issue). More recently, attempts have been made to include the uncertainty and its persistence dimensions into the discount rate discussion that seems to dramatically increase the expected net present value of future payoffs [57]. We may conclude that in any calculation of TEV as defined in equation (6), we have to choose the appropriate relevant period of time, discount rate, and method of discounting in order to reflect intergenerational preferences and uncertainty into the valuation process.

4. So .... just what is ecosystem TEV useful for?

A clear answer to this question suggests it be broken down into other two questions. Firstly, given the existence of controversy towards theoretical, methodological, and empirical aspects of ecosystem TEV, what is the usefulness of a monetary measure for ecosystem conservation decisions? Secondly, where the economic value is a theoretical, abstract measure experiencing a somewhat exacerbated controversy, what is the point of estimating it and using it for conservation issues? Let us begin with the first.

When government classifies some region as a protected area, biological resources of that region may be put under threat because the responsibility of management is transferred from people who live inside or close to the protected area to governmental agencies, located far away from the region. However, the direct costs of protection typically fall on the inhabitants who otherwise might have benefited from exploiting the Nature. Thus, there is a problem of justice and of intergenerational equity, which may have severe consequences if inhabitants have an economic disadvantage and, as a consequence, environmental regulations may be easily evaded or avoided.

National governments and local administration always faced difficulties in protecting their natural areas because the implementation and management of protection and conservation programmes are expensive tasks. Although society has been demonstrating the value it places upon intergenerational environmental transfers, by accepting to bear the opportunity cost of setting aside land and resources in their natural or almost natural state only for scientific, educational, recreation, and biodiversity protection or sustainability purposes, protected areas generally have being face budget problems, and insufficient personnel. The aforementioned financial shortages, together with rising economic pressure over environment, are on the alert to the necessity of adopting new mechanisms to improve Nature protection sustainability. Economists, biologists and management conservationists they all increasingly recognise the urgent need for alternative approaches to Nature management to introduce incentives or stimulus specifically used to incite and/or motivate economic, social and administrative agents (the stakeholders) to comply with Nature conservation strategies, to divert resources (e. g. land, capital, and labour) towards Nature conservation, and to incentive the participation of those groups and agents that work in or live within protected areas, to adopt sustainable development options.

Market-based incentives (including ecosystem valuation) are considered as the more adequate to perform the task. The basic underlying idea is that critic Nature has to be incorporated into the market system because Nature exhibit quasi-pure public good characteristics, an open access's free rider problem, together with lack or insufficiently stated property rights [12,17,58-59]. Because of these market failures individual users have too little incentive to conserve species and ecosystems because they earn immediate, high monetary benefits from exploiting Nature, without paying the full social and economic costs of its depletion. Instead, individual users transfer these costs to society to be paid either now or in the future. Therefore, although users may benefit collectively from managing Nature in a sustainable way on a protected area based conservation management scheme, from the individual point of view he/she can be better off by cheating and increasing their own use. As for the property rights problem it often happens that over exploited environmental resource and landscape tend to be the ones with the weakest or even non-existent ownership. However, in more developed countries like some Europeans, it often happens that biodiversity has private owners. But in the absence of markets, the private owner of the natural resource is not enabled to capture the benefits of Nature conservation. Hence, the private rational option will be to exploit the resources to extinction or its irreversible destruction. This bunch of market failures and its negative consequences upon environment explain why an effective and sustained government intervention is required to meet the values and needs society expects to have from conservation, all together with market forces [12-13, 19]. By using market-based incentives like fees, rewards, fines, compensations for ecosystem services, subsidies, and loans (or credit), land banks, revolving funds or daily wages, financing stimuli, voluntary action stimuli, and involvement of government authorities will be more incentivised. Therefore an increasing of compliance with the Nature conservation legislation is to be expected: firstly, because they may give local communities the financial means to develop sustainable activities; secondly because they must reduce economic pressure on environmental rich lands, by incentive locals to devote them instead to biological conservation and to concentrate economic and social activities on less biological rich lands; thirdly because they must conserve traditional knowledge and cultural systems; and fourthly because they must compensate locals for possible foregone income associated with the protection option.

As for the ecosystem's economic valuation, many economists and non-economists consider it as a very important conservation tool, an incentive measure, a support for conservation decision-making, and one of the ways to ensure that the private profitability gap between sustainable and unsustainable use of ecosystem services is narrowed or even closed [2, 21]. They see ecosystem valuation as a support for improving political and judicial decision-making within contexts of accruing environmental degradation, especially when compounded by increasing social demand for environmental services. Ecosystem valuation can also play a beneficial role in government land use, conservation, and tax planning, particularly when there is a growing interest in incentive or compelled conservation by private owners and property developers [60]. The basic economic decision methodology more commonly used to make cost-effective choices between different investment alternatives, the Cost-Benefit analysis, has often been under-utilised (or non-utilised at all) for environmental purposes because the monetary value of ecosystems is not took into account as an important variable. The absence of recognition towards the monetary value of ecosystem's and the services they provide to society is a common flaw of the decision process which happens because the existence of that value itself is not easily recognised by the stakeholders; besides, ecosystem's valuation is very demanding from the technical and the financial points of view. Further, economic ecosystem's valuation is important, as economic conservation instruments based on cash payments or tax breaks require the estimation of enough compensation to offset habitat losses elsewhere. Additionally, evaluation will be particularly important whenever ecological assets are not traded, or when it is necessary to make a comparison between ecosystems with different characteristics and, therefore, different values (ecosystem's value depends crucially on: the ecosystem location; its relationship with human activities and expected ecosystem changes over time), bringing confidence to ecosystem trading schemes in terms of the maximisation of net social benefits [61].

Strong theoretical, empirical, and jurisdictional progress has been made in the last three decades in the USA and Europe [62-64, 2], concerning the estimation and appliance of the economic values estimations to Nature conservation. Recently, one the more important efforts to improve the use of ecosystem valuation as a conservative management tool, was borne at the Heiligendamm Summit on 6-8 June 2007, where the G8 countries and the 5 major leaders newly industrializing countries endorsed a proposal suggested by the German government during the meeting that took place in Potsdamin on 20 March 2007 to make a study to analyze the global economic benefit of biological diversity, the costs of the loss of biodiversity and the failure to take protective measures versus the costs of effective conservation (the Potsdam Initiative for biodiversity). The study called The Economics of Ecosysems and Biodiversity (TEEB), was appointed to Pavan Sukhdev, and the preparatory work was initiated by the German Federal Ministry for the Environment and the European Commission, with the support of several other partners; hosted by UNEP, several countries and organizations have joined TEEB. Several reports were already produced (http://www.teebweb.org/InformationMaterial/TEEBReports/tabid/1278/Default.aspx) concerning the evaluation of costs of the loss and decline of Nature worldwide and their comparison with the costs of effective conservation and sustainable use. The main issue of the study, consisted in making visible the value of ecosystems and services to facilitate development of cost-effective policies [22].

There isn't, however, a one-hundred per cent consensus about the relevance of using ecosystem valuation as an incentive tool in conservation policies. Some economists argue that evaluation is neither necessary nor sufficient for conservation defending the key theme of conservation as making this policy more attractive than any alternative usage through translating the social benefits of ecosystems into income [17]. In their opinion, this does not necessarily oblige evaluation. However, one can question this opinion by asking how one can translate those ecosystem social benefits into income efficiently, in the absence of markets, invisibility of the ecosystem's monetary values, within contexts of growing financial resource scarcity, social equity promotion, and efficiency concerns? It clearly seems that valuation is the foundation to achieve that issue. Further, economic valuation constitutes a useful tool for educating and involving local populations and stakeholders by highlighting the connection between the ecosystems' underlying biophysical properties and benefits associated with the active or passive use of its services. It enables the justification of conservation projects encouraging local inhabitants and stakeholders to accept and to comply. If local people are conveniently alerted to the true economic value of their land, including scarcity, they will anticipate a higher future value and hold back from economic activities that may not be compatible with conservation. This behaviour will be smoothly endogenised by local inhabitants and stakeholders where accompanied by decisions that transform conservation decisions into sustainable income for local communities. In sum, environmental valuation can be seen as an economic conservation instrument that promotes ethics and fairness in conservation policies. Through environmental valuation and improved economic compensation mechanisms, society will reinforce its right to defend social ecosystem services even when such a defence may sometimes imposes severe economic and social restrictions on the use of land that belongs to other people independent of their social-economic development expectations.

Being such an important instrument for convincing people to comply and participate in conservation policy, there remains ongoing debate and a persistent reluctance as to the reliability, and validity of the monetary measures like the one defined by equation (6). Philosophers, ecologists and even some economists not subscribing to the idea of using environmental economic values as conservation tools have their own ethical and philosophical, technical and methodological arguments. It is unquestionable that ecosystem's TEV is an abstract, theoretical measure and does not measure the real value (but what is really the real value of something, one may questioning) of the thing as commonly understood. Economists tend to structure their thinking around models of perfect rational agents, with fixed preferences, making decisions in order to maximise certain well defined individual objectives, and this is very different from the paradigms characterising the natural sciences or even other social sciences (see [65] where both economic and ecological meaning of value, are compared). As a consequence of this hyper-abstract economic model of thinking, economic value is a *relative* not an *absolute* measure. It is the answer to a question involving a choice: how much an individual is willing to pay (to accept) for a unit more (or to forego) of an environmental resource without changing his or her current (future) wellbeing. To say one thing has greater economic value than another is an alternative way of saying that, under the circumstances, this would be chosen in preference to that. A correct understanding of what the economic value concept really means helps in understanding the value paradox (the paradox value is one of the arguments used by psychologists to demonstrate the incoherence of the economic monetary measure and its valuation shortcomings): articles that command great prices are often things that common people consider as being of little intrinsic or useful value, take diamonds or paintings for example. On the other hand, absolutely essential goods are often available at a negligible, or even no price, like water or landscape, at least in regions where such natural resources are abundant. There is, however, no inconsistency where goods that are, overall, immeasurably useful being worth less than the non-essential as economists determine marginal values: a good is worth more than another when the individual is not yet satiated and it is more difficult to obtain an additional unit than another unit of something else. People generally consider a diamond much more valuable than a litre of water, even while knowing the latter's vital importance, because they recognise the rarity of diamonds and the relative abundance of water. In short, the economic money measure of value basically reflects the scarcity of the good being valued under certain circumstances but not value from a common sense perspective. The economic money measure of value also reflects important issues that affect scarcity as perceived by the individual such as the existence of substitutes as well as the context surrounding the object of choice. In short, one may say that the theoretical, abstract weaknesses of economic valuation, identified by some commentators, is simultaneously its strength when used to improve decisions involving scarce non-market transacted resources, like ecosystems. Thus, while the level of abstraction and technical rigour of the theory underlying the definition of economic value are seen by some as handicaps, others consider them as trump-cards when there is a need for credible support to political and judicial decisions. However, these are complex methodological problems and the high level of economic and econometric expertise needed to estimate ecosystem values lies at the source of another set of criticisms. It is also a very expensive process. Nevertheless, these technical, methodological, or budgetary contras must not be used as arguments to turn our backs on ecosystem valuation although they must be taken into account during the political decision-making process when conservation decisions involve very important ecosystems at the local, national, or regional levels.

In short one must not agree with the argument any number is better than no number, used by some proponents, as it denotes an indefensible, very resigned attitude. When talking about economic values, economists are not referring to *any* number but to *a* number rigorously and theoretically defined and carefully applied in order to capture the value people put on the object of choice under certain circumstances of choice. Rather than saying any number is better than no number is to say an economic number is better as a minimum reference number, than any number at all.

5. Conclusions

In this chapter, we have discussed the advantages and disadvantages of ecosystem economic valuation as an incentive measure for enforcing local community co-operation in conservation decisions and management. The utilitarian approach allows value to arise in a number of ways depending on individual use of ecosystems. Hence, economists have generally settled for taxonomy of total ecosystem value interpreted as Total Economic Value (TEV) that distinguishes between Direct Use Values and Passive (Non-use) Values. TEV is a relative value and an answer mostly expressed in monetary terms to a carefully defined question in which two alternatives are compared. This answer depends on elements of choice defining the prevailing context. As ecosystems are not purchased on markets, one has to elicit individual preferences directly by the use of questionnaires such as Contingent Valuation (CV) to assess the individual's WTP and WTA relevant monetary measures. Complex criteria and "rules of evidence" such as those suggested by the NOAA's Panel must be applied to guarantee the reliability and validity of the CV data. To calculate TEV based on individual CV data and the asset analogy, time and uncertainty have to be considered when discounting service flows. One concludes that TEV may be a useful tool as an incentive, a support for decision-making, and as a tool for education and information. The fact of being a very abstract instrument, and very demanding from the theoretical and technical points of view, becomes an advantage. To date, it is still the only existing, carefully defined and applied and somehow reliable way of society knowing how much an ecosystem is worth within a market-based scenario.

Author details

Isabel Mendes
ISEG, Department of Economics /SOCIUS/CIRIUS, Technical University of Lisbon, Portugal

6. References

[1] IUCN (1994) Guidelines for Protected Area Management Categories. Gland, Switzerland: World Commission on Protected Areas.

[2] OECD (1999) Handbook of Incentive Measures for Biodiversity . Paris: OECD.

[3] Dudley, N., Stolton, S., Belokurov, A., Krueger, L., Lopoukhine, N., MacKinnon, K., Sandwith, T. and Sekhran, N. editors (2010) Natural Solutions: Protected areas helping people cope with climate change. IUCN-WCPA, The Nature Conservancy. Gland, Switzerland and Washington D.C.: UNDP, Wildlife Conservation Society, The World Bank and WWF.

[4] Kettunen, M., Bassi, S., Gantioler, S., and ten Brink, P. (2009) Assessing Socio-economic Benefits of Natura 2000 – a Toolkit for Practitioners. Brussels, Belgium: Output of the European Commission project Financing Natura 2000: Cost estimate and benefits of Natura 2000, IEEP.

[5] Mulongoy, K.J. and Gidda, S. B. (2008) The Value of Nature: Ecological, Economic, Cultural and Social Benefits of Protected Areas. Montreal: Secretariat of the CBD.

[6] Chape, S, Spalding, M., and Jenkins, M., editors (2008) The World's Protected Areas. Status, Value, and Prospects in the 21st Century. London: UNEP-WCMC, University of California Press.

[7] Balmford, A and Whitten T. (2003) Who Should Pay for Tropical Conservation, and How Could the Costs Be Met? Oryx 37, no. 02: 238-250.

[8] Borrini-Feyerabend, G., Kothani, A. and Oviedo, G. (2004) Indigenous and Local Communities and Protected Areas. Towards Equity and Enhanced Conservation. UK: IUCN, Gland, Switzerland and Cambridge.

[9] Goodstein E. S. (1999). Economics and the Environment. New Jersey: Prentice-Hall.

[10] OECD. (1996) Préserver la Biodiversité Biologique. Les Incitations Economiques. Paris: OECD.

[11] Wells M. and Brandon K. (1992) People and Park. Linking Protected Area Management With Local Community. (Washington DC: World Bank.

[12] McNeely J.A. (1988) Economics and Biological Diversity: Developing and Using Economic Incentives to Conserve Biological Resources. Gland: IUCN.

[13] Constanza R., Cumberland J., Daily H., Goodland R., and Norgaard R. (1997) An Introduction to Ecological Economics. St Lucie: International Society for Ecological Economics.

[14] Gerlagh R. and Keyzer M. (2002) Efficiency and Conservationist Measures: an Optimist Viewpoin. Journal of Environmental Economics and Management, 46: 310-333.

[15] Serageldin I. (1996). Sustainability and the Wealth of Nations. First Steps in an Ongoing Journey. ESD Studies and Monographs Series nº5 . Washington D. C.: World Bank.

[16] Jakobsson K.M. and Dragun A.K. (1996). Contingent Valuation and Endangered Species. Methodological Issues and Applications. Cheltenham: Elgar.

[17] Mendes I. (2003). Pricing Recreation Use of National Parks for More Efficient Nature Conservation: an Application to the Portuguese Case. European Environment 13: 288-302.

[18] Heal G. (2000). Nature and the Marketplace: Capturing the Value of Ecosystem Services. Washington D. C.: Island Press.

[19] Metrick A. and Weitzman M. (2000) Conflicts and Choices in Biodiversity Preservation. In R.N. Stavins editor. Economics of the Environment. New York: W. W. Norton & Company.

[20] Smith V.K. (2002) JEEM and Non-Market Valuation: 1974-1998. Journal of Environmental Economics and Management 39: 351-374.

[21] TEEB. (2008) The Economics of Ecosystems and Biodiversity – interim report. The European Commission. Available: http://ec.europa.eu/environment/nature/biodiversity/economics/pdf/teeb_report.pdf. Accessed 2012 May 4.

[22] Goulder L.H. and Kennedy D. (1997) Valuing Ecosystems Services: Philosophical Bases and Empirical Methods. In G.C. Daily editor. Nature's Services. Societal Dependence on Natural Ecosystems. Washington D. C.: Island Press: pp. 23-47.

[23] Turner R.K. (1999) The Place of Economic Values in Environmental Valuation. In I.J. Bateman and K.G. Willis editors. Valuing Environmental Preferences. New York: Oxford University Press: pp. 17-42.

[24] Carson R.T., Flores N.E. and Mitchell R.C. (1999) The Theory and Measurement of Passive-Use Value. In I.J. Bateman and K.G. Willis editors. Valuing Environmental Preferences. New York: Oxford University Press: pp. 97-130.

[25] Randall A. (1991) Total and Non-use Values. In J.B. Braden and C.D. Kolstad editors. Measuring the Demand for Environmental Quality. Amsterdam: North-Holland: 303-322.

[26] Smith V.K. (1993) Non-market Valuation of Environmental Resources: an Interpretative Appraisal. In R.N. Stavins editor. Economics of the Environment. New York: W. W. Norton & Company: 219-252.

[27] TEEB (2011). The Economics of Ecosystems and Biodiversity for National and International Policy Making. UK: Earthscan.

[28] Daily, G. C. editor (1997) Nature's services. Societal dependence on natural ecosystems. Washington, DC: Island Press: 392p.

[29] Chichilnisky G. and Heal G. (1998) Economic Returns From the Biosphere. Nature 391: February.

[30] Hicks J.R. (1939) The Foundations of Welfare Economics. Economic Journal 49: 696-712.

[31] Kaldor N. (1939) Welfare Propositions of Economics and Interpersonal Comparisons of Utility. Economic Journal 49:549-552.

[32] Kopp R.J. and Pease K.A. (1997) Contingent Valuation: Economics, Law and Politics. In R.J. Kopp, W.W. Pommerehne and N. Schwarz editors. Determining the Value of Non-Marketed Goods. Boston: Kluwer Academic Publishers: 7-59.

[33] Mäler K.G. (1971) A Method of Estimating Social Benefits from Pollution Control. Swedish Journal of Economics 73:121-133.

[34] Mäler K.G. (1974) Environmental Economics: a Theoretical Inquiry. Baltimore: Johns Hopkins University Press.

[35] Hicks J.R. (1943) The Four Consumer Surpluses. Review of Economic Studies 11: 31-41.

[36] Hanemann W.M. (1999) The Economic Theory of WTP and WTA. In I.J. Bateman and K.G. Willis editors. Valuing Environmental Preferences. New York: Oxford University Press: 42-95.

[37] Hanemann W.M. (1991) Willingness to Pay and Willingness to Accept: How Much Can they Differ?. In W.E. Oates The Economics of the Environment. Cambridge: University Press.

[38] Freeman A.M. III. (1993) The Measurement of Environmental and Resource Values: Theory and Methods. Washington DC: Resources for the Future).

[39] Mitchell R.C. and Carson R.T. (1989) Using Surveys to Value Public Goods: the Contingent Valuation Method. Washington DC: Resources for the Future.

[40] Alberini, A. and Khan, J.R. (2009) Handbook on Contingent Valuation. Edward Elgar: Cheltenham.

[41] Kettunen, M.; Bassi, S.; Gantioler, S. and ten Brink, P. (2009) Assessing Socio-economic Benefits of Natura 2000 – a Toolkit for Practitioners November 2009 Edition. Output of the European Commission project Financing Natura 2000: Cost estimate and benefits of Natura 2000. IEEP, Brussels, Belgium. Available: http://ec.europa.eu/environment/nature/natura2000/financing/docs/benefits_toolkit.pdf. Accessed 2012 May 07.

[42] Braden J. and Kolstad D. editors (1991) Measuring the Demand for Environmental Quality. The Neetherlands: North-Holland.

[43] Hufschmidt M.M., James D.E., Meister A.D., Bower B.T., and Dixon J.A. 1(983) Environment, Natural Systems and Development. Baltimore: The Johns Hopkins University Press.

[44] Bateman, I. and K. Willis (1999). Valuing Environmental Preferences: Theory and Practice of the Contingent Valuation Method in the US, EU, and Developing Countries. Oxford: Oxford University Press.

[45] Arrow K., Solow R., Portney P.R., Leaner E.E., Radner R., and Schuman H. (1993) Report of the NOAA Panel on Contingent Valuation. *58 Federal Regulation 4601 et seg.*

[46] Green C. and Tunstall S. (1999) A Psychological Perspective. In I.J. Bateman and K.G. Willis. Environmental Preferences. New York: Oxford University Press: 207-257.

[47] Fischoff B. and Furby L. (1988) Measuring Values: a Conceptual Framework for Interpreting Transactions with Special Preference to Contingent Valuation of Visibility. Journal of Risk and Uncertainty 1: 147-184.

[48] Fischhoff, B. (1991) Value Elicitation: is there anything in there? *American Psychologist* 46: 835-47.

[49] Rosenberger R.S., Peterson G.L., Clarke A., and Brown T.C. (2003) Measuring Dispositions for Lexicographic Preferences of Environmental Goods: Integrating Economics, Psychology and Ethics. Ecological Economics, 44:63-76.

[50] Spash C.L. (2000) Ecosystems, Contingent Valuation and Ethics: the Case of Wetland Re- creation. Ecological Economics, 34: 195-215.

[51] Batabyal A., Kahn J.R., and O'Neill RV. (2003) On the Scarcity Values of Ecosystem Services. Journal of Environmental Economics and Management, 46: 334-352.

[52] Baumol W.J. (1968) On the Social Rate Discount. The American Economic Review, 58: 788-802.

[53] Ramsey F. (1928). A Mathematical Theory of Saving. Economic Journal, 38: 543-559.

[54] Harrod R. (1948). Towards a Dynamic Economics. London:.MacMillan.

[55] Heal G.M. (1993). The Optimal Use of Exhaustible Resources. In: A.V. Kneese and J.L. Sweeney editors. Handbook of Natural Resource and Energy Economics. Amsterdam: North-Holland.

[56] Bazerlon C. and Smetters K. (1999). Discounting Inside the Beltray. Journal of Economic Perspectives, 13: 213-228.

[57] Newell R.G. and Pizer W.A. (2003). Discounting the Distant Future: How Much do Uncertain Rates Increase Solutions? Journal of Environmental and Management, 46:52-71.

[58] Hanemann M. (1988). Economics and the Preservation of Biodiversity. In: Wilson E. O. editor. Biodiversity. Washington DC: National Academic Press: 193-199.

[59] Randall A. (1988). What Mainstream Economists Have to Say About the Value of Biodiversity. In: Wilson E. O. editor. Biodiversity.: Washington DC.: National Academy Press: 217-223.

[60] Boyd J. and Wainger L. (2003). Measuring Ecosystem Service Benefits: the Use of Landscape Analysis to Evaluate Environmental Trades and Compensation", *Discussion Paper 02-63, Resources for the Future.* Available: *http://www.rff.org/rff/Document/RFF-DP-02-63.pdf*. Accessed 2012 May 08.

[61] McGarthland A. and Oates W. (1995). Marketable Permits for the Prevention of Environmental Deterioration. Journal of Environmental Economics and Management, 207: 207-228.

[62] Bonnieux F. and Rainelli P. (1999). Contingent Valuation Methodology and the EU Institutional Framework. In: I.J. Bateman and K.G.Willis editors. Valuing Environmental Preferences. New York: Oxford University Press: 585-612.

[63] Loomis J.B. (1999). Contingent Valuation Methodology and the US Institutional Framework. In: I.J. Bateman and K.G. Willis Valuing Environmental Preferences. New York: Oxford University Press: 613-628.

[64] Atinkson, G. and Mourato, S. (2008). Environmental Cost-Benefit Analysis. Annual Review of Environmental Resources 33: pp 317-344.

[65] Farber S.C., Constanza R., and Wilson M.A. (2002). Economic and Ecological Concepts for Valuing Ecosystems Services. Ecological Economics, 41: 375 – 392.